FUNDAMENTALS OF WORK MEASUREMENT

What Every Engineer Should Know

FUNDAMENTALS OF WORK MEASUREMENT

What Every Engineer Should Know

Anil Mital
Anoop Desai
Aashi Mital

CRC Press
Taylor & Francis Group
Boca Raton London New York

CRC Press is an imprint of the
Taylor & Francis Group, an **informa** business

CRC Press
Taylor & Francis Group
6000 Broken Sound Parkway NW, Suite 300
Boca Raton, FL 33487-2742

Printed on acid-free paper
Version Date: 20160815

International Standard Book Number-13: 978-1-4987-4582-6 (Paperback)

Visit the Taylor & Francis Web site at
http://www.taylorandfrancis.com

and the CRC Press Web site at
http://www.crcpress.com

Table of Contents

Preface

This is not a textbook on Work Measurement for Industrial Engineers (IEs). Rather, this book is intended to be a brief guide to nonindustrial engineers, business managers, and those responsible for undertaking or supervising industrial and service activities. It aims at providing basic information on work measurement techniques and procedures so that individuals who are not trained as industrial engineers may not only understand these techniques but also, on occasion, may use them as well. It should help them understand the engineering and scientific principles associated with measuring work.

Why is such a book necessary when there are many excellent texts on work measurement? Many engineers, who are not necessarily IEs, need to know how to measure work in the course of their routine work. For instance, mechanical engineers involved in design work, manufacturing engineers designing for manufacturing, and costing engineers engaged in various aspects of cost estimations. These individuals do not have an IE background but require it in the course of their work. This book is intended to be a simple resource that can educate them in the basics of work measurement. Even IEs, who take work measurement courses as a part of their basic background knowledge buildup, can benefit from a simple primer on work measurement if they have strayed away from the field for a significant period of time.

This book is divided into 11 chapters. Chapter 1 introduces the concept of work and discusses the importance on measuring it. Chapter 2 which can be omitted without the loss of basic information on work measurement, deals with techniques and procedures that help in reducing the excess work content, and thus the inefficiencies, of an activity; that is job or work design. Since work or job design is primarily an IE activity, this chapter may be used for background purposes.

Chapters 3 through 7 describe various aspects of work measurement. In Chapter 8, we focus on white-collar work, primarily in the service industry. In Chapter 9, we describe work measurement techniques that do not rely on time as a measure of work. Work measurement instruments and some of the associated software are described in Chapter 10. Finally, in Chapter 11, we discuss how work standards help estimating costs particularly in jobs that are predominantly manual in nature.

The book was motivated by years of teaching mechanical, manufacturing, electrical, and other nonindustrial engineers who take courses in the area of design for manufacturing and manufacturing cost estimation. These individuals need work measurement information in order to become proficient in techniques associated with designing for maintenance, assembly, disassembly, functionality, etc., but have neither taken courses in work measurement or ergonomics nor have any such course a part of their formal coursework. Further, while they may have an interest in knowing how to measure work, they are unlikely to pick up a text on work measurement written primarily for IEs and mostly focusing on work design.

We hope that this brief introduction to work measurement for non-IEs will be useful to individuals who need to know how to measure work without getting into the complexities of work or job design. We are also hopeful that this guide will also benefit those who are closely associated with industrial and service work.

Finally, we wish to thank our publisher for appreciating the need for such a book in spite of the existence of many excellent books on Work Design and Work Study.

Anil Mital

Anoop Desai

Aashi Mital

Authors

Anil Mital is a former professor of mechanical engineering and manufacturing engineering and design at the University of Cincinnati. He is also a former professor and director of industrial engineering and a professor of physical medicine and rehabilitation at the University of Cincinnati. He earned a BE in mechanical engineering from Allahabad University, India, and an MS and PhD in industrial engineering from Kansas State University and Texas Tech University, respectively. He is the founding editor-in-chief emeritus of the *International Journal of Industrial Ergonomics*, and also the founding editor-in-chief emeritus of the *International Journal of Industrial Engineering*. He is also the former executive editor of the *International Journal of Human Resource Management and Development* and author/coauthor/editor of over 500 technical publications, including 25 books and over 200 peer-reviewed archival journal articles. Some of his books have been translated into Dutch, Spanish, and Korean languages. His recent research interests include application of DFX principles to product design, economic justification, manufacturing planning and facilities design, and design and analysis of human-centered manufacturing systems. He is the founder of the International Society (formerly Foundation) of Occupational Ergonomics and Safety and winner of its first Distinguished Accomplishment Award (1993). He is a Fellow of the Human Factors and Ergonomics Society and recipient of its Paul M. Fitts Education Award (1996) and Jack A. Kraft Innovator Award (2012). He is also a recipient of the Liberty Mutual Insurance Company Best Research Paper Award (1994) and the American Association for Engineering Education's Eugene Grant Award (1988). A Fellow of the Institute of Industrial Engineers, Dr. Mital received its distinguished Dr. David F. Baker Award for lifetime research activities and for advancing the discipline of industrial engineering. Dr. Mital has also received the Ralph R. Teetor Award from the Society of Automotive Engineers. He is also one of Elsevier Science Publishers Most Cited Authors.

Anoop Desai is an associate professor of mechanical engineering in the College of Engineering and Information Technology at Georgia Southern University. He earned his PhD in industrial and manufacturing engineering from the University of Cincinnati in 2006. His primary research interests are product life cycle management and design. His research deals extensively with design for "X" principles, focusing on green design, environmentally conscious manufacturing, and design for maintainability. He is also actively involved in research and teaching in the areas of engineering economy, new product development, CIMS, and quality control. He has authored over 70 technical articles, including 25 archival journal articles, and his work is widely cited.

Aashi Mital earned an MA in Nineteenth-Century American History and Classical Archaeology from the University of Cincinnati in 2014. Her research and publication interests focus on Received Memory in Post-Civil War Southern Culture, Radical Reconstruction, the Industrial Revolution, and Geopolitical Relations of the Greek Archaic Period. She is a professional consultant and vice-president of the popular Museums and Historic Sites of Greater Cincinnati colloquium. Mital strives to create parallels, by means of interdisciplinary work, in order to reexamine historical methodologies and create new approaches in today's multifaceted industry. She has been a copy and language editor for the *International Journal of Industrial Engineering: Theory, Applications and Practice* for nearly a decade. Along with contributing to other prominent journals and articles on human productivity to the industrial engineering literature, Mital continues to publish on groundbreaking historical and archaeological research, historical forecasting, and political sustainability. She also coauthored the well-known engineering text *Product Development: A Structured Approach to Consumer Product Development, Design, and Manufacture.*

1 Introduction: Significance of Work Measurement

1.1 WHAT IS WORK?

Work is defined as purposeful or intentional activity that is undertaken, or physical or mental effort that is exerted by an individual in order to accomplish something. This "something" could be making objects or accomplishing tasks of varying difficulties. Therefore, cooking, writing, fetching water, painting, doing craft work, building infrastructure (houses, bridges, etc.), making furniture or artifacts, sewing garments, shaping materials to achieve specific objectives, etc. are all examples of purposeful activities that are defined as "work."

To accomplish any task or activity, often referred to as "work" or "jobs," effort is required. This effort is typically measured in terms of time it takes to complete the activity or the work contained within it (sometimes this effort may be measured in terms of time a person can sustain an activity—endurance time). There are two components to the work that is contained in any activity:

1. The basic work content of the activity
2. Additional, or excess, work content

The basic work content is defined as the minimum effort as measured by the time it takes to complete an activity. This amount of time is the theoretical minimum time required to produce the object of interest or complete the process and cannot be reduced. Conditions must be perfect so that the required effort is the "minimum" possible. In reality, such an occurrence would be unlikely if not impossible. Even if it was possible to eliminate all excess work content, it may not always be desirable to do so for various reasons, including the ability to respond to unexpected breakdowns or prescribed maintenance.

The activity completion time that is required in practice is increased due to the inclusion of additional work content. This additional work content is the result of inefficiencies introduced by factors as follows:

1. Poor design of the product
2. Poor choice of the process
3. Poor choice of material or its utilization
4. Poor utilization of the primary and support processes
5. The worker and the management

In practice, the increase in work content (time) introduced by these inefficiencies may be reduced to approach the basic work content (theoretical minimum time required) but can never be achieved. Figure 1.1 shows the basic and additional work content of completing an activity or making an object. The following section briefly discusses certain key factors that add to the basic work content.

1.2 FACTORS ADDING WORK CONTENT TO A PRODUCT (OR ACTIVITY)

1.2.1 Product Design

A product is intended to provide function(s) that the buyers require and features that they desire. Providing these functions and features at a price that people are willing to pay and the reliability of the function a product is expected to provide are critical requirements that a product must achieve. A product with numerous components made up of diverse materials and relying on fasteners, such

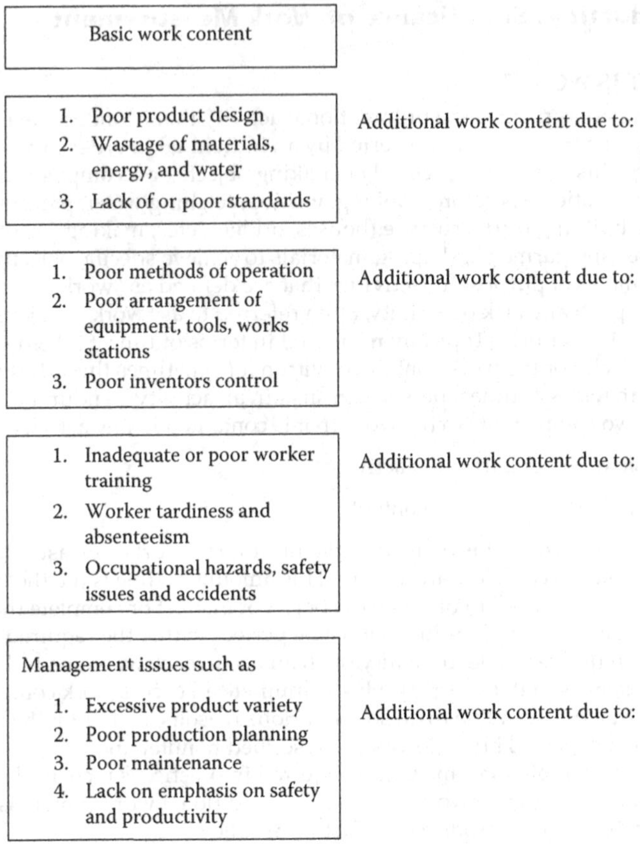

Figure 1.1 Basic and additional, or excess, work content.

as screws, leads to increased work content. As a general rule, the number and the types of materials used in a product should be minimized. Instead of using fasteners, such as screws and nuts or bolts, it is recommended to use snap-in type of fasteners.

On the other hand, a product with too few components increases the complexity of components and tends to reduce its reliability (failure to provide the function when needed). This eventually leads to increased product cost, thus making it economically less desirable. A good product design provides a balance between cost and reliability.

There are also considerations such as handling and the orientation of the components over the course of the product manufacture. Parts that are easy to grasp have high feedability (can be oriented in a number of acceptable ways due to the symmetry or only in a very specific way due to extreme asymmetry), and are self-locating to reduce the work content.

1.2.2 Lack of Standardization and Incorrect Quality Standards

A large number of nonstandard parts not only increase the assembly time but also lead to inventory problems (storing multiple parts, several vendors, and supply chain issues). Nonstandard parts lead to increased costs as such parts cannot be easily procured (off-the-shelf vs specially made parts). Components should be standard or standardized as much as possible to take advantage of the economy of the scale. This aids in managing the supply chain by allowing many vendors vs a few (too few vendors can create the possibility of supply chain shocks).

The quality standards are critical as well. If the quality standards are too high (narrow tolerances), then additional processing of materials may be required, thus adding to the work content. If the process is not in control, then this could lead to many parts being rejected, as they would fall outside the acceptable regions. Narrow tolerances, on the other hand, could lead to fewer parts containing defects and thereby increasing reliability. Too loose quality standards (wide tolerances) could result in large number of defective products and parts with low reliability. It is worth mentioning that tight control on processes has made it possible to achieve high quality standards and increased parts' reliability in recent years. This, however, has not solved the problem of material removal, like that of machining, to achieve tight quality control. Therefore, choosing the right quality standard is essential in controlling the work content.

1.2.3 Poor Choice of Process or Method of Operation

Poor choice of manufacturing process, methods of handling or maintenance, inefficient method of work can all add to the work content. For instance, using machining instead of injection molding or net- or near-net shape casting can add to the processing time. Use of inappropriate material handling equipment, such as using forklifts instead of conveyor belt to move materials continuously, adds to the total time as does the use of poor preventive maintenance policies that lead to frequent equipment breakdown and work stoppage.

Furthermore, work methods that are tedious and cumbersome add to the excess time. Consequently, all methods must be analyzed for improvement and reduction of nonproductive time.

Equally important are issues that deal with the utilization of space. Proper planning of the facility to ensure smooth flow of materials is critical. This saves time and money by reducing effort.

1.2.4 Reducing Waste

Reducing waste, be it of materials, energy, water, or labor, is crucial to reducing the work content. Such expenditures are a reflection on the additional work content. Products, processes, and activities must be analyzed to ensure that minimal amounts of materials and supplies, including those of water, energy, and labor, are utilized. Additionally, materials that are removed during manufacture are recycled.

Designing products for "end-of-life" is the norm in modern enterprises. A "product recovery" approach needs to be utilized at the end of the life of a product. The recovery must be at the product level, module level, part level, and material level. The options could be

Repair—restore to working condition

Refurbishing—improve to quality level, though not like new

Remanufacturing—restore to quality level, as new

Cannibalization—limited recovery

Recycling—reuse materials only

The objective is to recover as much as possible the economic as well as the ecological value of products, components, and materials, in order to minimize the ultimate quantity of waste. If a component is repaired and reused, the ultimate work content gets distributed over a larger quantity.

The argument applies to the conservation of energy and water as well. Not all sources of energy are equally economical—energy from natural gas may be cheaper than energy from oil in some locations. The use of economical sources of energy, energy-efficient equipment, tools, and processes is an important consideration.

Reducing excess wastage of water in processes, recycling water, and replacing processes that are water intensive are also critical. Not only can water be expensive, but it may also be a scarce resource in areas suffering from drought or areas that have high population density.

Methods that are inefficient and waste energy or water are examples of poor product design and excess work content.

1.2.5 Poor Management and Poor Worker Performance

Examples of poor management are

- Poorly designed and poorly performing products
- Too many products, poor ecological value of products and too many varieties
- Poor maintenance
- Poor utilization of space, equipment, and tools
- Poor production planning and control
- Inadequate safety and poor health policies

Some of the causes of poor worker performance are

- Inadequate and improper training
- Lack of discipline and supervisory control, causing work delays, etc.
- Absenteeism and tardiness
- Lack of responsibility

These factors, directly or indirectly, influence the work content and need to be controlled or mitigated. Our intent is not to discuss how to accomplish that but to highlight how the work content and thereby the completion time are affected by these factors.

1.3 WHAT IS WORK MEASUREMENT?

In the most basic form, work measurement is defined as the application of techniques that determine the time it takes a qualified worker to carry out a task at a determined rate of working. In other words, work measurement is the determination of the duration a task takes (time/task). The British Standards Institution established this definition of work measurement in 1991 (British Standard 3138).

Measurement of time/task, however, is not the only way to measure work. For instance, work can be measured by determining the metabolic energy expenditure rate of an individual performing a task (oxygen consumption which is a reflection on the task energy requirement). Such methodologies are discussed in Chapter 9.

Despite alternative or additional methods available to measure work, the determination of how long it takes to complete a task is widely accepted as the definition of work measurement.

Work measurement is one of the two components of what constitutes "work study":

1. Method study
2. Work measurement

While the work measurement part of work study is aimed at establishing the duration of a task, the method study focuses on task simplification and efficient methods of doing it. Thus, work study requires that jobs be designed for efficiency first and then measured (time/task) for the work content. Even though Chapter 2 does discuss the techniques that help critically examine tasks from the standpoint of eliminating inefficiencies and making simplifications and improvements, our focus is on the work measurement aspect of work study.

A major goal of industrial engineers is to improve work efficiency. This is accomplished by conducting methods study. Unless we can reduce the excess work content as much as possible, there is little point in measuring work. Thus, method study and work measurement go hand in hand.

While the reduction in the total work content falls primarily within the domain of industrial engineers, many disciplines are interested in measuring work. This book, therefore, focuses on work measurement.

1.4 WHY IS IT IMPORTANT TO MEASURE WORK?

As previously stated, work measurement is the application of techniques that help us determine the time a qualified worker would take to complete a specified task when working at a defined speed (rate of working). The time that is established by the management for performing a designated task or job is called the "work standard" and although time is the only consideration in these standards, the term "work standard" is considered all encompassing.

The most apparent question one would ask at this point is, "Why are these work standards important?" The general answer is that work standards help us control labor performance and allow us to extract the best each worker has to offer. This, in turn, helps us improve productivity and the standard of living. Yet, control of labor performance is not the only way to enhance productivity—maximizing the utilization of resources is a major contributor to productivity enhancement.

Productivity is generally defined as given below

$$\text{Productivity} = \frac{\text{Outputs}}{\text{Inputs}}$$

where **Outputs** are defined as the products and/or services rendered by an enterprise and **Inputs** are the resources utilized in rendering those outputs. Typically, the inputs are

- Assets (capital, land, building, etc.)
- Materials (raw materials, supplies, etc.)
- Equipment, facilities, and tools
- Labor (direct and indirect)
- Energy
- Water

Water is an important resource due to its global scarcity and the usual definitions of productivity do not include it in the list of inputs. Yet, the very nature of its scarcity, increasing global population, and changing climatic conditions and its use in many industrial processes make it an extremely important resource. Take, for instance, the use of water in the process of fracking.

Increased productivity means increased output from the same quantity of inputs or decreased amounts of inputs for the same output. In other words, increased productivity is the result of greater control of wastage. It is this wastage that leads to excess work content. The method study component of work study allows us to systematically record and analyze our use of resources (inputs) in order to make improvements. Thus, while work measurement helps us establish the work standards (time/object), it alone cannot help us enhance productivity by simply controlling labor performance. While our focus in this book is on work measurement, it should be kept in mind that both method study and work measurement (also referred to as time study) are needed to make productivity improvements.

Work measurement, which leads to work standards, has many uses. The following is a list of some of the major uses (these are not in any particular order):

- Design and evaluation of product designs
- Design and evaluation of efficient equipment
- Selection of equipment, tools, jigs, and fixtures
- Process and operations planning
- Production scheduling (shop loading and work schedules)
- Facilities design (plant layout and materials handling)
- Budgets and cost controls
- Labor cost estimations
- Setting selling price
- Evaluation of labor performance
- Production line balancing
- Establishing a fair day's work and wages
- Implementing incentive schemes
- Job evaluation
- Collective bargaining
- Increase employee morale by providing an objective method of evaluation
- Evaluate the effect of training methods and learning
- Determining manpower requirements
- Establishing production capacities
- Making a determination regarding plant expansion (physical expansion vs adding a shift)

This is not a comprehensive list, as many more items could be added. What is clear is that work measurement has a number of critical and important uses. Given below is a brief discussion of some of the factors listed above.

1.4.1 Product Design

Basic product design includes inputs from a number of sources—design engineers, materials and manufacturing engineers, sales and marketing, quality control department—and must provide the needed function at an acceptable cost. Further, the design must lead to a desirable and useful product. During the course of developing products, a number of designs are developed. However, only one of these designs, or a composite thereof, can get to the final production. Work standards provide an objective method for evaluating these different designs on the basis of costs by

- Providing a cost comparison of various methods of manufacture (different methods require different amounts of effort, for example, machining vs casting)
- Providing a cost comparison with different materials (some materials are more expensive than others and may require a different process)
- Providing a cost comparison owing to different levels of quality that can be accomplished with different processes and materials
- Providing a cost comparison of designs by comparing the required assembly time
- Providing a cost comparison of designs by comparing maintenance requirements, etc.

Since product design process is complex and requires multiple and often conflicting requirements, the above list of design comparisons can be easily extended to include design criteria such as functionality and usability.

1.4.2 Equipment, Tool, Fixture, and Jig Design

Work standards allow comparison of different equipment, tools, etc. by allowing comparison of

- Cost of manufacture of items in different categories
- Cost of alternative methods of manufacture and materials for different items in each category
- Cost of various design options for each item in each category
- How each design would influence the method of operation, maintenance requirements, quality achieved, etc.
- The utility and energy requirements of various items in each category, etc.

1.4.3 Selection of Equipment and Tools

Work standards allow the selection of appropriate equipment by permitting a comparison of the cost of setup, cost of operation, production capacity, cost of maintenance, reliability, etc.

1.4.4 Processing and Operations Planning

Work standards allow determination of the plant production capacity by allowing determination of

- Time/task for each activity
- Additional cost that will be incurred by alternative methods that may be necessary if the output has to be increased

- Operation split (perform an operation on one machine vs multiple machines; single cuts vs multiple cuts, etc.)
- Line balancing (how often the work will advance from stage to stage)
- What operation(s) will be performed in what order
- What different assembly activities will take place at each station, the number of workstations, etc.

1.4.5 Production Scheduling

Production scheduling requires determination of

- Workload for each machine and worker
- Workload for each workstation
- Workload for each department
- Scheduling start and finish of work orders and shipping dates, etc.

Work standards are essential in order to accomplish these efficiently.

1.4.6 Labor Requirement

Work standards allow the determination of manpower requirements. Knowing time/piece can help determine the work capacity of individuals. Once the daily output needs are known, it is simple to calculate how many people would be needed. Work standards also help determine the training requirements (by allowing skill comparisons), training schedules, and training effectiveness. They also permit comparison of different training methods and training schemes.

1.4.7 Wages and Wage Incentives

Work standards permit the determination of what a fair day's work is and thus how much workers should earn. Work standards are also essential in establishing wage incentives as these allow determination of productive work rates (above and beyond acceptable day's work work rates).

1.4.8 Comparison of Work Methods and Task Evaluations

Work standards are useful when comparing two or more methods of doing the same thing. Clearly, a method that takes less time is more desirable as it is more productive. Work standards also help compare different jobs. The amount of time spent in different jobs provides a measure of the effort required to perform different jobs. The same concept can be applied to different aspects of a job, providing information about which elements of a job require more effort than the others.

1.4.9 Preparing Budget, Controlling Costs, and Establishing Selling Price

Work standards provide an essential tool for cost control by permitting a comparison between actual and standard performances. It provides the basis for estimating direct and indirect labor costs. By providing labor cost for operations, it provides information for preparing budgets, direct and indirect costs/unit or/hour or/workstation, etc.

Knowing the costs, which are a function of factors, such as production volume, size, and materials, one can set the selling price. While there are many variables that affect the selling price, it cannot be set without knowing how much the product costs—work standards are essential for that.

1.4.10 Facilities Design

It involves the following:

- The physical layout of the plant (location of departments, equipment, etc.)
- The material handling system

Work standards allow determination of information such as

- The number of machines required
- How many machines a worker can handle
- Which material handling methods are more desirable
- The type of layout (product vs process)
- Type of material handling equipment
- Single vs multiple floors
- The pattern of material flow within the plant
- Placement of equipment and line balancing, etc.

This information, in conjunction with information about operations and processes, and scheduling, form the backbone of a productive enterprise.

1.5 SUMMARY

In this chapter, we have defined work and looked at some factors that contribute to nonproductive work. We have also defined work measurement and its importance in improving productivity. We have also reviewed the various uses of work measurement. The relationship between work measurement and work study and methods study has also been discussed.

Some of the topics discussed, such as product design and facilities design, are quite involved and have had only a cursory discussion in this chapter. Although Industrial Engineers may have more in-depth background in these areas, it is not within the expertise areas of other engineering and business disciplines. Therefore, we have provided a list of selected references at the end of this book under the heading "Suggested Reading" for those who wish to further their background in these areas.

2 Prior to Measuring Work: Minimizing Inefficiencies

Nations that prosper are better able to provide for their citizens' basic needs and help them improve their quality of life. Once the basic needs of people are met (food, clothing, shelter, security, and health), they aspire for things that may be categorized as luxuries. Items such as the quality of clothing, size of housing, cultural needs, enjoyment of leisure time, and personal hobbies fall in this category. The ability of people to achieve these luxuries depends on their standard of living. Prosperity and the living standard are thus tied—greater the prosperity of a society, higher its standard of living and vice versa.

For a nation to be able to raise its standard of living, it must improve its productivity (defined in Chapter 1). Increased productivity leads to economic growth and a higher standard of living and better quality of life. Productivity improvements can be achieved in one of the two following ways:

- Increase quality and quantity of output from the same amount of resources
- For the same output, reduce the resource inputs by changing the quality of inputs (e.g., better quality machines, materials, etc.)

Work study aims at increasing productivity by recording tasks and activities and systematically examining the ways in which the excess work content (nonproductive work content) of tasks and activities can be reduced or minimized. This is followed by developing work performance standards. The segment of work study that deals with recording and examining tasks and activities is called the methods study (as we defined in Chapter 1, methods study aims at improving resource utilization by improving ways in which things are done).

Thus, the goal of methods study is to permit "smart" or "efficient" working by reducing excess work content, not "hard" or "ineffective" or "laborious" working by working "fast." Significant higher productivity gains can be achieved by working smart than working hard.

While, all jobs and activities must be subjected to the critical examination of methods study eventually, the process cannot begin by randomly selecting a task. The order in which the selection of jobs should take place should depend on how much to the cost or profit a job or activity contributes. The following section discusses how to choose the job/work/task/activity to study.

2.1 SELECTING JOB OR ACTIVITY TO BE STUDIED

As stated above, any job can be selected for a critical examination with an aim to improve its method of operation. Selecting jobs randomly, however, is not a very productive way of making productivity gains. The job selection efforts should focus on the following:

- Economic considerations such as
  - Which jobs are most expensive?
  - Which products have the highest volume?
  - Which products contribute most to the profits?
  - Which operation is most labor intensive?
  - Which operation has most repetitive work? And so on
- Technological considerations such as
  - Which machines are most unreliable?

- Which technologies are old and should be replaced with new ones?
- Which process or what materials should be replaced?
- Which operations require most energy? And so on

■ Worker-related considerations

- Which jobs or activities are most monotonous?
- Which jobs or activities are most dissatisfying?
- Which jobs or activities have most safety problems?
- Which jobs or activities can benefit from additional training or education? And so on

One of the methods that may be used to identify tasks or operations that should be subjected to the methods study focus first is the Pareto analysis (named after an Italian economist). This method is also known as the ABC analysis or value analysis. This method allows one to focus efforts on jobs that are important, instead of wasting time on unimportant things.

Figure 2.1 shows the concept of Pareto analysis or Pareto distribution, as it is often called. Basically, the Pareto distribution states that in any enterprise, a small number of items account for the bulk of costs or profits (criterion—y-axis in Figure 2.1). The criterion could be something else besides profit or cost, for example, waste, energy consumption, storage needs, breakdowns, and accidents.

As Figure 2.1 shows "A" items account for 10% of the total number of parts but account for 70% of costs or profits (or resources or whatever the criterion might be). "B" items account for 25% of the parts and 25% of the costs or resources.

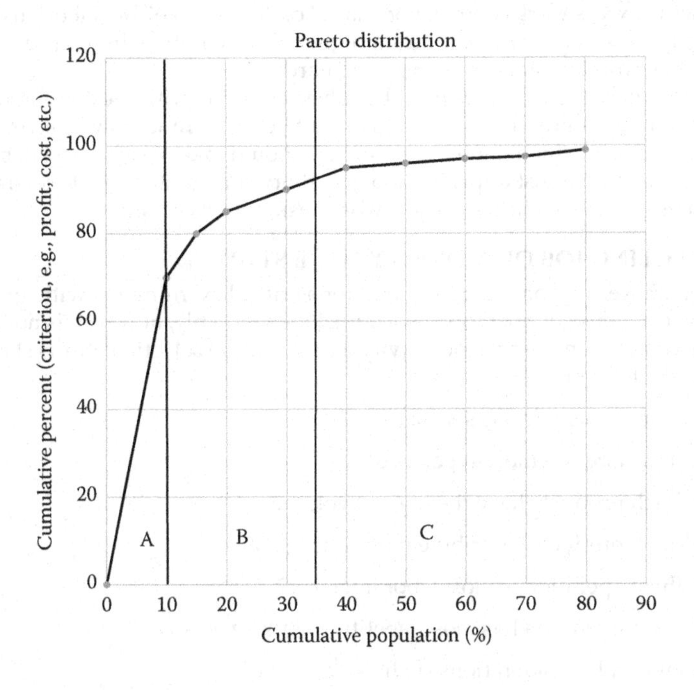

Figure 2.1 Pareto distribution.

"A" and "B" parts together are about one-third of all parts but account for the bulk (95%) of the costs. The remaining bulk (65%) of the parts accounts for just 5% of the costs.

The intent here is to focus on "A" parts or jobs or activities first, followed by "B" parts, jobs or activities in order to make the biggest impact on excess work reduction and productivity enhancement, rather than get bogged down in dealing with the "C" category. If one has to deal with product design issues or quality issues or reliability issues or direct labor issues or any cost, design or profit issues, the focus needs to be on the "A" list of parts, products, jobs, or activities. For instance, few parts in a product are most responsible for its quality and the focus, therefore, ought to be on those parts if the intent is to improve a product's quality.

One may also look at the Pareto distribution as a histogram or frequency occurrences. Preparation of a Pareto distribution is simple. For example, if one wants to prepare a distribution of a company's profit contribution (criterion) by product, the following two steps need to be done. First, a table of all products and contribution to profit by each product needs to be developed. Next, the products need to be ranked by profit, from high to low. Typically, one would observe that a very small number of products contribute the most to the profit; these are "A" products identified in Figure 2.1. These products, jobs, or activities are the priority for the methods study.

2.2 RECORDING THE EXISTING WORK METHOD

Once a job or activity has been selected for methods study, everything that happens during the course of completing that task must be recorded *in writing*. The emphasis must be *only on facts* that are related to the existing method, as it is used to complete the work associated with the selected task. This written record permits critical examination and analysis of the facts at the later stages.

Writing what happens in a narrative is not an effective method as the processes used in industry are complex. Further, a narrative does not accurately convey what is really happening; it not only takes a considerable amount of space, but it is subjective as well. Therefore, it is essential to rely on tools that convey what the method actually is. Charts and diagrams have been developed as tools to record methods. These specialized charts and diagrams are listed in Table 2.1. Not all charts and diagrams are widely used. Some of the most widely used charts are process charts. The following subsections discuss the most widely used charts.

The charts listed in Table 2.1 are of two kinds:

- Charts that indicate the sequence of the process
- Charts that use time scale

Diagrams, on the other hand, indicate movement and provide a graphical representation of information. Thus, diagrams are a supplement to charts.

The recording of information on a process chart is done with the use of five symbols. These symbols are

Indicating operation

Operation indicates a main or major step in the method, procedure, or process. This symbol is typically used when some change or modification takes place.

Table 2.1: Common Charts and Diagrams Used to Record Information

Charts Indicating Sequence of Actions
- Outline process chart
- Flow process chart
  - Person type
  - Material type
  - Equipment type
- Two-handed process chart
- Procedure flowchart

Charts Indicating Use of Time Scale
- Multiple activity chart
- Gang chart
- SIMO (simultaneous motion cycle) chart

Diagrams Showing Movement
- Flow diagram (2-D or 3-D)
- String diagram
- Cyclegraph
- Chronocyclegraph
- Travel chart

For instance, the change in the shape of a part or a product undergoes during machining or assembly is an *operation*.

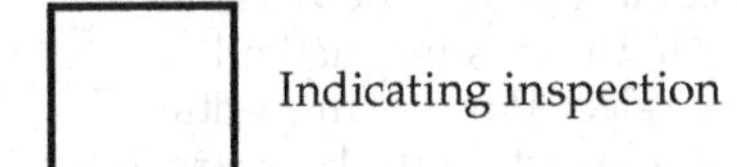

Indicating inspection

Unlike operation, inspection does not indicate a change or modification or change of state. It is, rather, something done to ensure quality, or a check or verification for something. Increasingly, manufacturing operations are being carried out without inspection due to control on process variations.

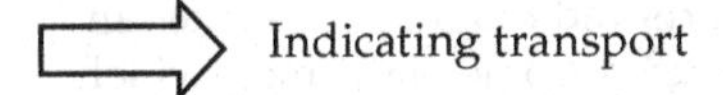

Indicating transport

When an object, part or product or equipment or worker, moves or is moved from one place to another, a *transport* occurs. It is commonly used during materials handling.

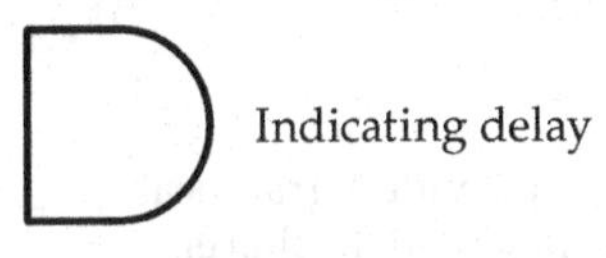

Indicating delay

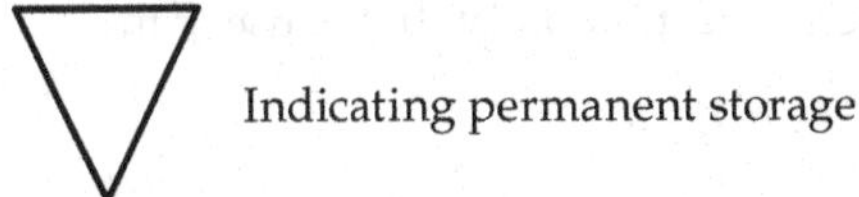

Indicating permanent storage

When material is put in a warehouse or storage, where it is protected and stays for prolonged periods of time (days and weeks instead of minutes and hours) and can only be removed by proper authorization, the *storage* is considered *permanent*. In general, items in permanent storage either are finished items, products, or raw materials. Work in process is rarely put in permanent storage—it may be subjected to delay, however.

Occasionally, it becomes necessary to show multiple activities occurring at the same time. For instance, an operation and inspection may happen simultaneously. In such instances, symbols for both activities can be combined by superimposing on each other—putting circle inside the square.

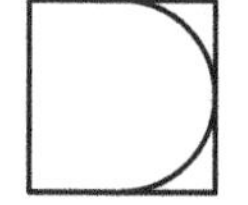

Indicating combined activities inspection and delay

Combined activities represent simultaneous activities or activities by the same operator at the same workstation. The combined symbol shown above may represent a delay in carrying out inspection; the delay necessitated perhaps due to workpiece preparation or curing period.

Although not considered a part of standard symbols described above, a diamond-shaped symbol, such as the one shown below, may be used to indicate decision-making.

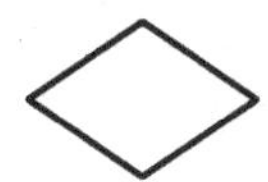

2.2.1 Outline Process Charts

The outline process chart is sometimes used when an overall picture, showing only the main operations and inspections, is required. This kind of chart also serves as a preliminary work prior to preparing a detailed flow process chart. Symbols for operation and inspection are the only symbols used in this case. The sequence of operations and inspections is shown along a vertical line. In case of multiple parts, such as those used in an assembly, multiple vertical lines (one for each part) may be used and assembly of the parts may be shown by using horizontal lines connecting vertical lines. It is convenient if operation and inspection time can be shown along with the operation and inspection number. Times are shown in hours.

Figure 2.2 shows the engineering drawing for cap-cylinder assembly. All dimensions are given in inches. The assembly has four parts

1. Teflon cap
2. Teflon tube
3. Machined cap
4. Machined tube

The following are the operations and inspections for each part.

Teflon cap	Operation 1	Setup machine (0.25 h)
	Operation 2	Turn 1.31 in. outside diameter (OD) (0.05 h)
	Operation 3	Turn 0.75 in. inside diameter (ID) (0.04 h)
	Operation 4	Face 2 concentric flanges (0.07 h)
	Operation 5	Cutoff (0.05 h)
	Inspection 1	Check all dimensions (0.07 h)
Teflon tube	Operation 6	Setup machine (0.50 h)
	Operation 7	Turn 1.31 in. OD (0.05 h)
	Operation 8	Turn 1.00 in. OD (0.09 h)
	Operation 9	Face end taper (0.05 h)

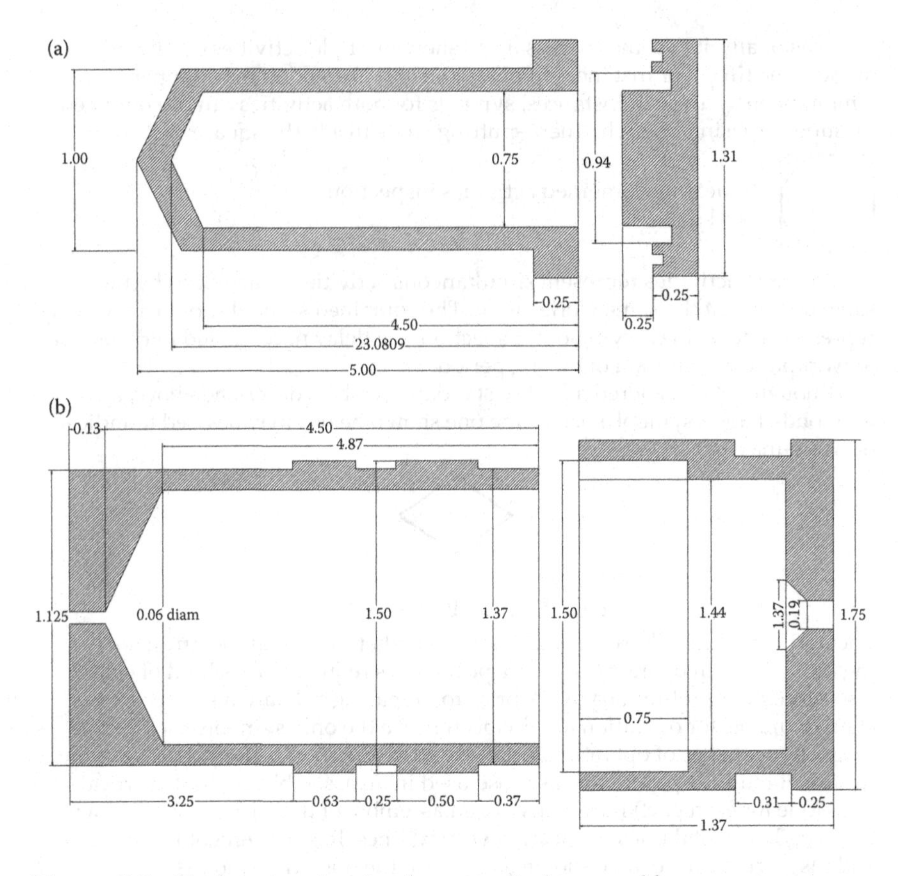

Figure 2.2 Engineering drawing for cap-cylinder assembly.

	Operation 10	Cutoff end adjacent to 1.31 in. OD (0.05 h)
	Operation 11	Bore 0.25 in. dia, pilot 4.75 in. deep (0.10 h)
	Operation 12	Bore 0.75 in. dia hole, 4.75 in. deep (0.09 h)
	Inspection 2	Check all dimensions (0.07 h)
Machined tube	Operation 13	Setup lathe (0.50 h)
	Operation 14	Turn 1.50 in. OD (0.09 h)
	Operation 15	Turn 2 1.375 in. OD flanges (0.08 h)
	Operation 16	Knurl 0.625 in. wide flange (0.06 h)
	Operation 17	Turn OD thread 0.500 in. long (0.05 h)
	Inspection 3	Check threads (0.07 h)
	Operation 18	Bore 0.25 in. dia pilot hole 4.875 in. deep (0.12 h)
	Operation 19	Bore 1.00 in. dia ID hole 4.875 in. deep (0.17 h)
	Operation 20	Turn 1.125 in. dia OD (0.18 h)
	Operation 21	Cutoff (0.04 h)
	Operation 22	Drill 1/16 in. dia hole (0.02 h)
	Operation 23	Deburr (0.03 h)
	Inspection 4	Check dimensions (0.07 h)

Machined cap	Operation 24	Setup lathe (0.50 h)
	Operation 25	Turn 1.75 in. dia OD (0.07 h)
	Operation 26	Knurl 0.25 in. wide flange (0.05 h)
	Operation 27	Knurl 0.94 in. wide flange (0.10 h)
	Operation 28	Turn 0.31 in. long flange, 0.06 in deep (0.06 h)
	Operation 29	Bore 0.50 in. dia pilot hole (0.06 h)
	Operation 30	Bore 1.44 in. ID to 1.37 in depth (0.19 h)
	Operation 31	Turn 0.75 in. long ID thread (0.05 h)
	Operation 32	Bore 1.50 in. ID (0.05 h)
	Operation 33	Cutoff (0.04 h)
	Inspection 5	Check dimensions (0.07 h)
	Operation 34	Drill 0.19 in. dia hole (0.02 h)
	Operation 35	Countersink (0.02 h)
	Operation 36	Deburr (0.03 h)
	Inspection 6	Check dimensions (0.07 h)
Assembly	Operation 37	Assemble Teflon tube and cap (0.02 h)
	Operation 38	Assemble Teflon tube/cap, machined tube (0.02 h)
	Operation 39	Assemble machined cap (0.02 h)
	Inspection 7	Final inspection (0.02 h)

Figure 2.3 shows the outline process chart for the cap-cylinder assembly.

2.2.2 Flow Process Chart

While the outline process chart gives an overall idea of the method or process, greater details are generally needed. This requires the construction of a flow process chart. A flow process chart shows the sequence, in detail, of the flow of a product or process; *all* events are recorded using the symbols described earlier.

As mentioned in Table 2.1, a flow process chart could be

- Worker type—recording what a worker does
- Equipment type—recording the use of equipment
- Material type—recording what happens to the material (its treatment and handling)

Figure 2.4 shows a general flow process chart blank form (a blank form is included as this is one of the most commonly used forms in methods study). It can be used for a worker or equipment or any material. The form is generic in nature and can be customized for convenience. It includes basic information such as the activity, total distance traveled, whether the method is current or proposed, etc.

Figure 2.5 shows how a flow process chart may be used. It shows cleaning of a gearbox. The overall activity involves picking up the gearbox, transporting it to the shop for cleaning, stripping the gearbox, cleaning and inspecting parts, taking parts to the degreaser in a basket for degreasing, and finally storing them.

This activity includes 5 operations, 11 transportations (loading, unloading, and movements), 2 delays, 1 inspection, and 1 storage. A total distance of 53 m is covered. All this information is based on *direct observation*.

As one can see, Figure 2.5 uses all five symbols described earlier, as opposed to the outline process chart (Figure 2.3) which uses only two (operation and inspection). It can also provide time information even though Figure 2.5 does not include time information.

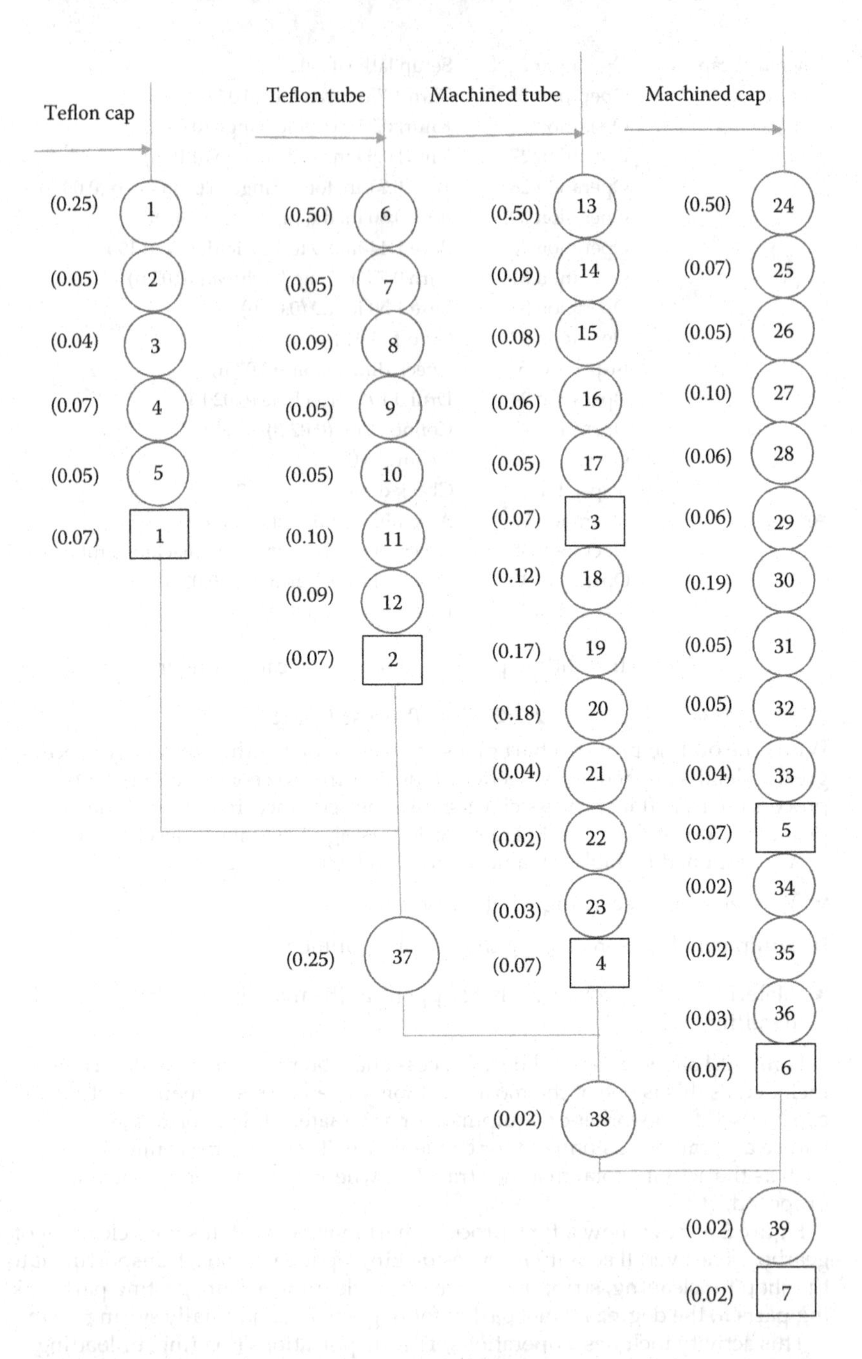

Figure 2.3 Outline process chart for the cap-cylinder assembly.

A major distinction between the flow process chart and the outline process chart is that while the outline process chart may include operational information about the whole product (assembly), a separate flow process chart may be necessary for each major component of the assembly.

Flow process chart	Material/Type			
Chart No. Sheet No. of	Summary			
Subject charted	Activity	Present	Proposed	Savings
Activity	Operation ● Transport ➡ Delay D Inspection ■ Storage ▼			
Method: Present/proposed	Distance (m)			
Location: Degreasing shop	Time (work-min)			
Operative(s): Clock Nos.	Cost			
Charted by: Date:	Labor			
	Material			
Approved by: Date:	Total			

Description				Symbol					Remarks
	Qty	Distance (m)	Time (min)	●	➡	D	■	▼	

Figure 2.4 Flow process chart form (blank).

Once all the observed information is recorded on the chart, including details about what is being recorded, who recorded it, when it was recorded, etc., the information is ready for critical examination to identify inefficiencies (excess work content).

The flow process chart shown in Figure 2.5 is "material" type—it shows the sequence of events happening to a piece of material. The same technique may be used to record the sequence of events a worker undertakes or a piece of equipment undergoes.

2.2.3 The Two-Handed Process Chart

This is a special kind of process chart and is typically used to record the sequence of workers working at a workstation. The workers at a workstation use both hands, hence the name two-handed process chart. The movements, or the lack thereof, of the two hands (and sometimes feet) are shown relative to each other. This type of chart is most useful when the movements are repetitive and cyclic in nature. The sequence of a complete cycle is recorded for analysis.

Each operation recorded on a flow process chart may have a number of activities. For instance, stripping the gearbox is recorded as a single operation in Figure 2.5. This operation most likely involves in number of steps. The

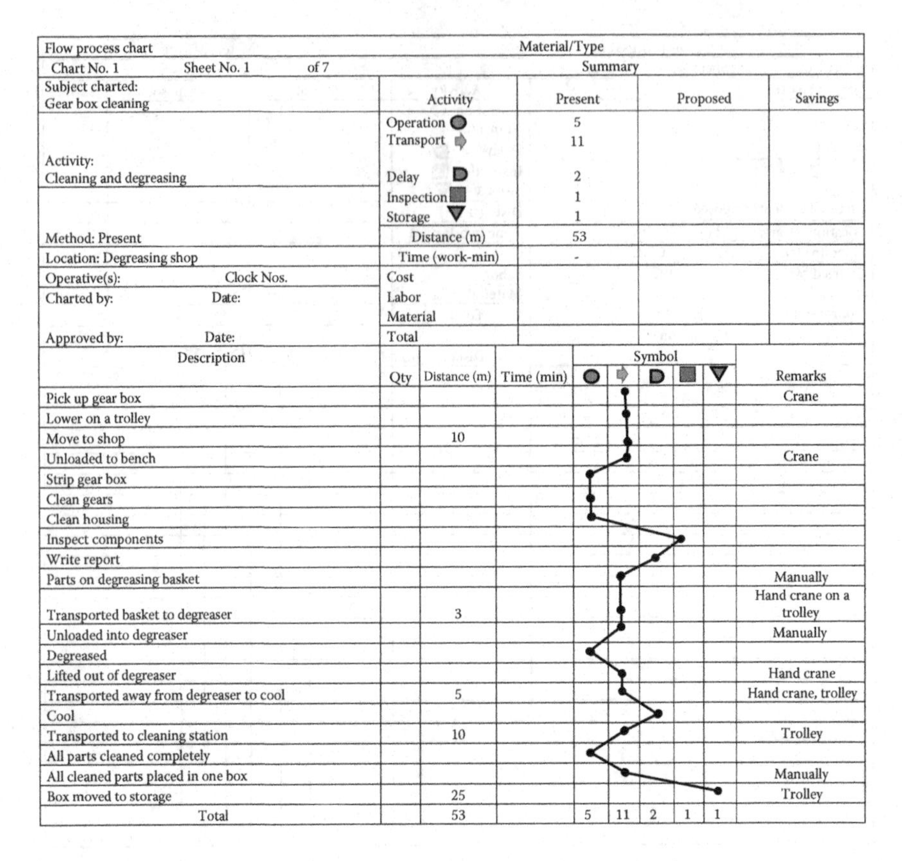

Flow process chart			Material/Type		
Chart No. 1	Sheet No. 1	of 7	Summary		
Subject charted: Gear box cleaning			Activity	Present	Proposed	Savings
Activity: Cleaning and degreasing			Operation ●	5		
			Transport ➡	11		
			Delay D	2		
			Inspection ■	1		
			Storage ▼	1		
Method: Present			Distance (m)	53		
Location: Degreasing shop			Time (work-min)	-		
Operative(s):	Clock Nos.		Cost			
Charted by:	Date:		Labor			
			Material			
Approved by:	Date:		Total			

Description	Qty	Distance (m)	Time (min)	●	➡	D	■	▼	Remarks
Pick up gear box					●				Crane
Lower on a trolley					●				
Move to shop		10			●				
Unloaded to bench					●				Crane
Strip gear box				●					
Clean gears				●					
Clean housing				●					
Inspect components							●		
Write report						●			
Parts on degreasing basket					●				Manually
Transported basket to degreaser		3			●				Hand crane on a trolley
Unloaded into degreaser					●				Manually
Degreased				●					
Lifted out of degreaser					●				Hand crane
Transported away from degreaser to cool		5			●				Hand crane, trolley
Cool						●			
Transported to cleaning station		10			●				Trolley
All parts cleaned completely				●					
All cleaned parts placed in one box					●				Manually
Box moved to storage		25						●	Trolley
Total		53		5	11	2	1	1	

Figure 2.5 Flow process chart (material type) for gearbox cleaning.

two-handed process chart is useful in recording the sequence of activities involved in stripping the gearbox. Thus, each operation recoded on a flow process chart could result in a two-handed process chart.

Only four of the symbols used in the flow process chart are used in the two-handed process chart. The symbol for inspection, a square, is not used (in the event an inspection is performed, for example, taking a closer look at the object being held, the symbol for operation is used). The four symbols that are used also have a different meaning:

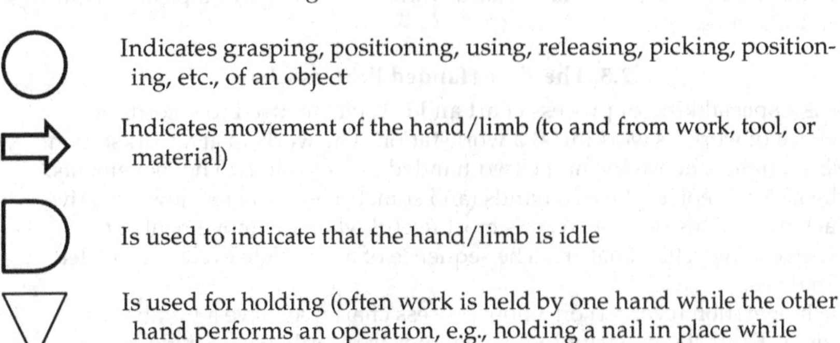

Indicates grasping, positioning, using, releasing, picking, positioning, etc., of an object

Indicates movement of the hand/limb (to and from work, tool, or material)

Is used to indicate that the hand/limb is idle

Is used for holding (often work is held by one hand while the other hand performs an operation, e.g., holding a nail in place while hammering)

Two handed process chart									
Chart No.	Sheet No.								Workplace layout
Location:									
Operative:									
Charted by:	Date:								
Left-hand description	●	➡	D	▼	●	➡	D	▼	Right-hand description

Summary				
Method	Present		Proposed	
	LH	RH	LH	RH
Operations				
Transports				
Delays				
Holds				
Inspections				
Totals				

Figure 2.6 Two-handed process chart (blank).

Using a two-handed process chart (Figure 2.6 shows a blank that may be customized) provides at least three major advantages:

- It allows detailed information about the activity, instead of simply recoding it as an operation, making it useful for in-depth examination
- It allows examination of each of the element that comprises the overall activity recorded as a single operation
- It permits recording a wide variety of activities at a workstation, such as machining, assembly, disassembly, and maintenance

The preparation of a two-handed process chart requires attention to some procedural details:

- Sequence for only one hand should be recorded at one time
- Multiple cycles should be studied to ensure that all elements of the operation are recorded correctly
- Actions that happen simultaneously should be recorded at the same level in the chart
- Actions occurring at different times should be recorded at different levels, above if happening before or below if happening later

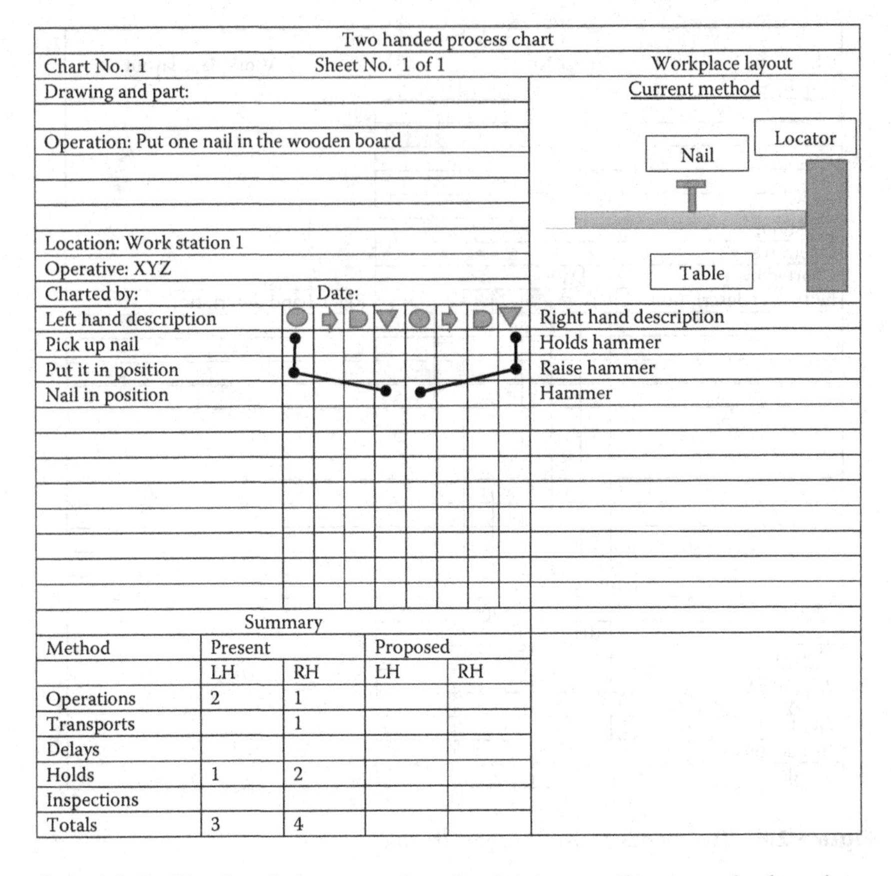

Two handed process chart

Chart No.: 1 Sheet No. 1 of 1

Drawing and part:

Operation: Put one nail in the wooden board

Location: Work station 1

Operative: XYZ

Charted by: Date:

Workplace layout

Current method

Left hand description	Right hand description
Pick up nail	Holds hammer
Put it in position	Raise hammer
Nail in position	Hammer

Summary

Method	Present		Proposed	
	LH	RH	LH	RH
Operations	2	1		
Transports		1		
Delays				
Holds	1	2		
Inspections				
Totals	3	4		

Figure 2.7 Two-handed process chart for driving a nail in a wooden board using a hammer.

- A sketch of the workplace layout should accompany the chart
- Combining operations should be avoided

Figure 2.7 shows a very simple example of driving a nail in a wooden board.

2.2.4 Procedure Flowchart

A procedure chart is used to record what happens to a document (a document type flow process chart) and is most useful in offices (of course, it can be used to record the sequence of shop floor paperwork as well). The chart records the sequence as a document moves from one department to another.

A columnar format is used as a document is typically routed from one department to another. The sequence of departments is from left to right. All activities taking place within a department are recorded in its column.

The greatest difficulty that arises in preparing a procedure chart is the lack of direct observation; the observer is unlikely to be present at the moment an activity takes place. Therefore, information is generally obtained by asking questions during an interview. The knack for eliciting information, therefore, is crucial and reflects on its accuracy.

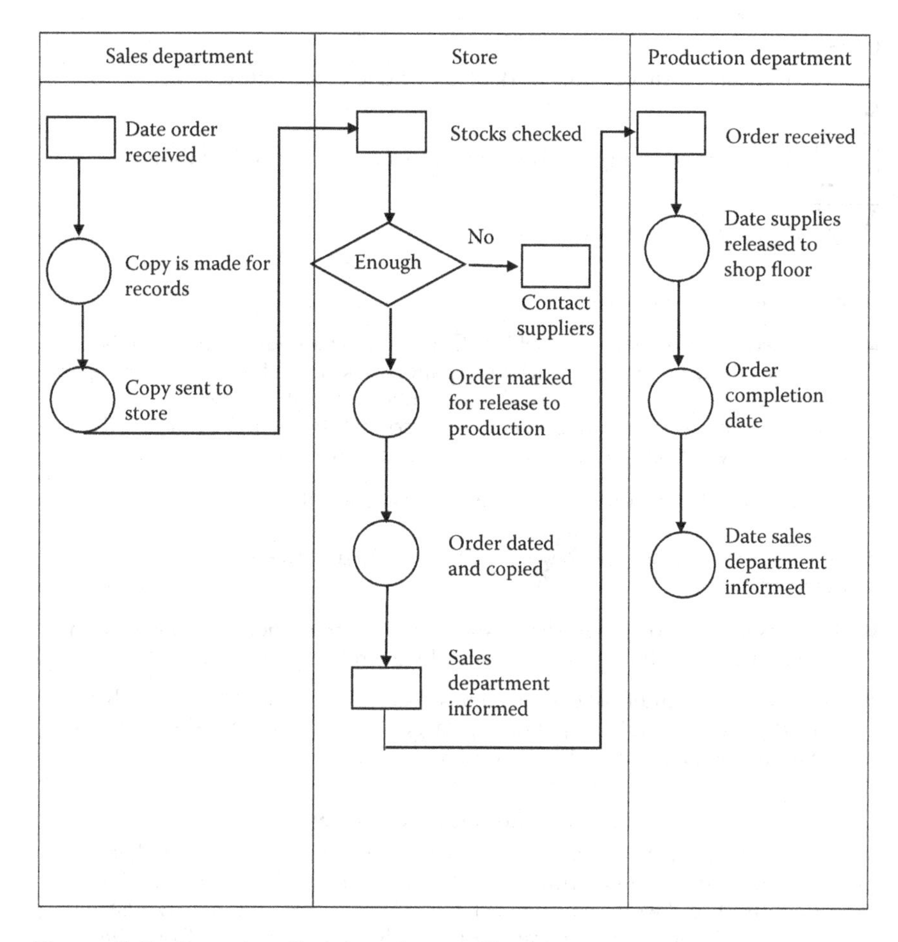

Figure 2.8 Procedure flowchart for an order form.

Figure 2.8 shows the flow of an order form as it moves from the sales department to storage to the production department.

Greater sophistication can be added to these charts by using various symbols. In addition to the five symbols used in the preparation of a flow process chart, two additional symbols described below may be used. The interpretation of the original symbols may also be modified slightly:

◎ Indicates creation, or origination, of a form, card, etc.

● Indicates addition of information or data to an existing form

○ Indicates handling operation, such as collation, folding, and retrieving

⇨ Indicates movement to a department

Indicates form or paperwork being held (e.g., signature, etc.)

Indicates verifying information, such as numbers and totals, etc.

Indicates actions such as filing

A few cautionary notes

- Increased paperwork is a fact of life. Considerable amount of time can be spent on filling, filing, and retrieving paperwork. Therefore, reducing paperwork is important. It is suggested that symbols described above should be used to prepare procedure process charts so that these can be examined for efficiency by reducing redundant or nonproductive work
- Delays should be recorded only when they are significant
- Special symbols, such as , may be used to describe shredding or destruction
- Symbols may be combined if necessary. For example, checking information and entering it on the form is indicated by a circle within a square
- Forms that have multiple carbon copies need to be charted separately. If possible, these then should be put on one large chart showing the separation of copies and sequence of individual copies

2.2.5 Multiple Activity Chart

Multiple activity charts record the activities of multiple workers or workers and machines against a common time scale. These charts are also known as "man–machine" charts or "work planning" charts. The activities of the worker(s) and machine(s) are indicated in separate vertical columns as "working" or "idle." The goal here is to identify ineffective (idle) time so that it can be reduced and utilization can be increased.

Maintenance and mass production are two activities that derive large benefit from using multiple activity charts. This is particularly true when machines have a large capital cost, or when operatives have large operating cost (hourly wages).

These charts are also helpful in determining the number of machines an operative, or a group of operatives, can manage in terms of operation and supervision.

The multiple activity chart records activities of worker(s) and machine(s) by recording "working time" and "idle time." The working time is indicated by a shaded area and the idle time by blank area (other conventions can also be used provided these are indicated on the chart). The recorded times for each operative and machine are plotted in the column associated with each operative and machine. The activities performed by each entity (worker or machine) should be indicated in the shaded area for identification.

Figure 2.9 shows a typical multiple activity chart for a group of four workers—some of the "workers" could be machines. Keep in mind that in recording time, it is not necessary to be extremely accurate—time may be recorded in multiple or fraction of hours or in total minutes; seconds are not necessary.

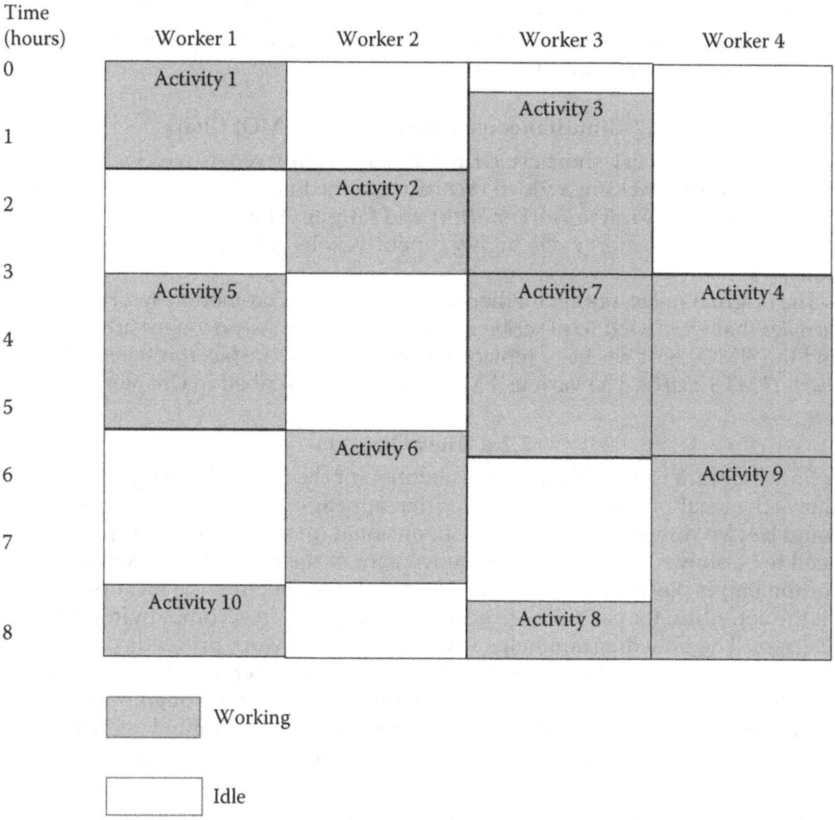

Time (hours)
Worker 1
Worker 2
Worker 3
Worker 4
0
Activity 1
1
Activity 3
2
Activity 2
3
Activity 5
Activity 7
Activity 4
4
5
Activity 6
6
Activity 9
7
Activity 10
8
Activity 8
Working
Idle

Figure 2.9 A typical multiple activity chart.

2.2.6 Gang Chart

There are situations where several workers work in a group (gang). Such situations are fairly common in construction, materials handling, and maintenance activities. Our interest here is to find out what each operative does and how much time is spent working and remaining idle each cycle. The multiple activity chart has been modified to record the sequence of activities in such situations. The result is the gang chart or the gang process chart.

The gang chart can be prepared in two ways:

- One way is to do it like the multiple activity chart, with one column devoted to each worker (Figure 2.9).
- The second way to use the format of the flow process chart, except the chart is prepared horizontally, devoting one row to each worker. The sequence of activities performed by each operator in a cycle is expressed using the flow process chart symbols (symbols are plotted horizontally). The first column has a listing of all the workers and the chart progresses from left to right. The vertical column indicates what each worker is doing at the same moment.

Both methods require taking an overview of what each worker is doing at different times and provide a snap of group's activities.

The chart is analyzed for delays with an aim to reduce idle time. Substantial savings can be achieved by rearranging activities or changing schedules so that the idle time (delays) is minimized.

2.2.7 Simultaneous Motion Cycle (SIMO) Chart

Some jobs have a very short cycle time and are highly repetitive. Such jobs are best recorded by making a video recording of the task. The aim in analyzing these jobs is to minimize worker effort and fatigue. The methods for recording and analyzing short cycle highly repetitive jobs are called *micromotion study*.

The original micromotion methodology relied on micromotion symbols called *therbligs* that were used to describe movements. These are no longer preferred and the SIMO chart has been replaced by *predetermined motion time standards charts (PMTS charts)*. The various PMTS charts are described in Chapter 6.

2.2.8 Flow Diagram

A flow diagram is a graphical representation of the plant (floor) layout and shows the location of all the activities that appear on a flow process chart. It could be two dimensional or three dimensional (in such cases multiple floors need to be shown). The path of the movement of the material (or worker or equipment) is plotted on the layout diagram by connecting the location of all the activities; for each flow process chart there is an accompanying flow diagram. The flow diagram helps visualize how the work proceeds on the shop floor as it moves from one activity to another. Together, the flow diagram and the flow process chart provide a complete picture of the ongoing activity. The activities on a flow diagram are identified by the symbol and by the number used in the flow process chart (operation 1, 2, 3,…; inspection 1, 2, 3,…, etc.).

Backtracking or repeated movements over the same path are shown by separate lines (or using a different color if computer graphics is used to prepare the flow diagram). Different colors may also be used if there are multiple materials or workers or equipment or a combination thereof.

Figure 2.10 shows a typical two-dimensional flow diagram, depicting material flow using flow process chart symbols (being unpacked, inspected, cleaned, marked, sprayed, and dried before being stored). Together with the flow process chart, the flow diagram provides a more complete and clear picture of the sequence of activities. This is particularly helpful in reducing travel distance, materials handling, and eliminating backtracking as changes can be made to the layout to make it more efficient.

2.2.9 String Diagram

Often the movement of workers takes place at irregular intervals. The movement may or may not include moving materials simultaneously. Further, the movements, while frequent, may not follow the same sequence in every cycle. Such movements occur both in manufacturing and nonmanufacturing environments. The following, for instance, are the commonly found examples:

- Workers supervising, loading or unloading multiple machines
- Removing and storing materials/products from continuous processes
- Work in textile and paper mills

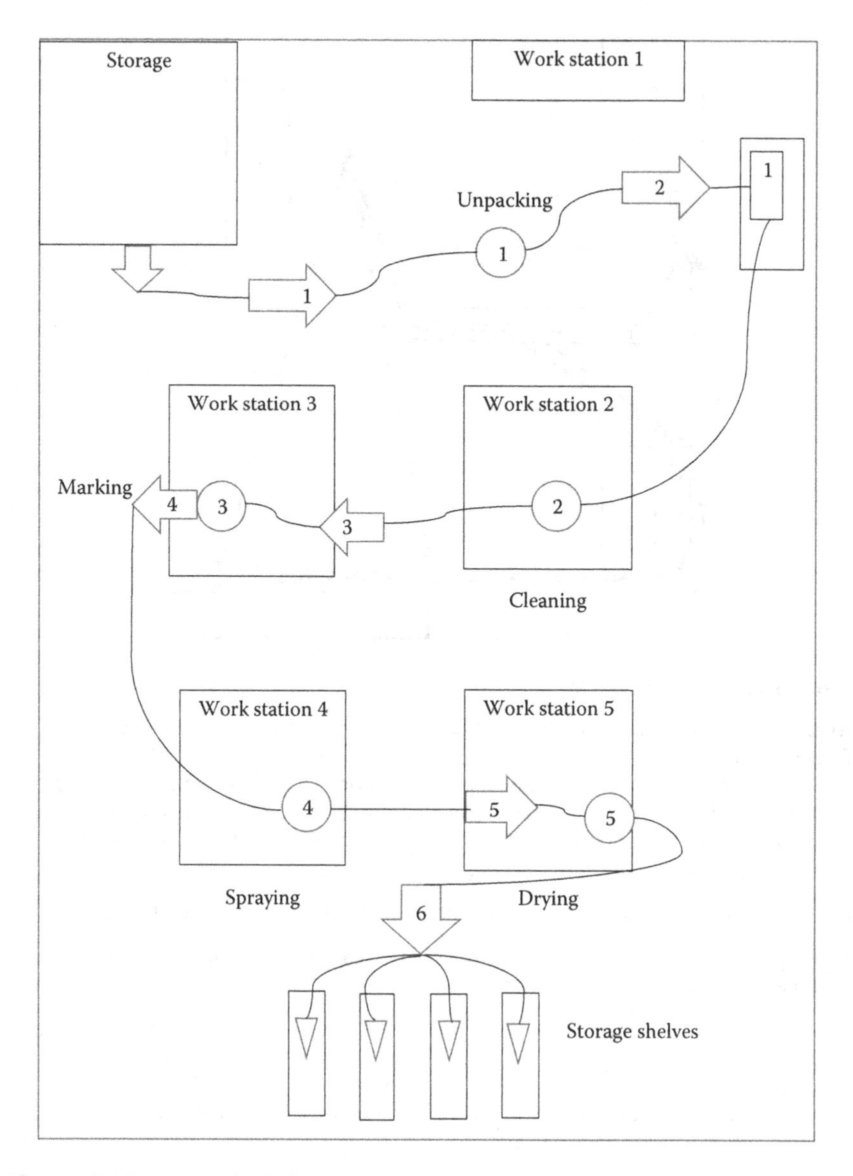

Figure 2.10 Typical 2-D flow diagram.

- Movements in storage areas (warehouses) and stores
- Movements in restaurants, mail rooms, and offices

Under such circumstances, string diagrams are used to study the movements.

A string diagram uses a thread, or string, on a scale model to track and trace the movements of workers, materials, or equipment during a sequence of events. The length of the string (thread) is a measure of the total distance traversed. The scale model, therefore, should be accurate. A string diagram can also be prepared to trace the movements of a worker's hands as they move from one point

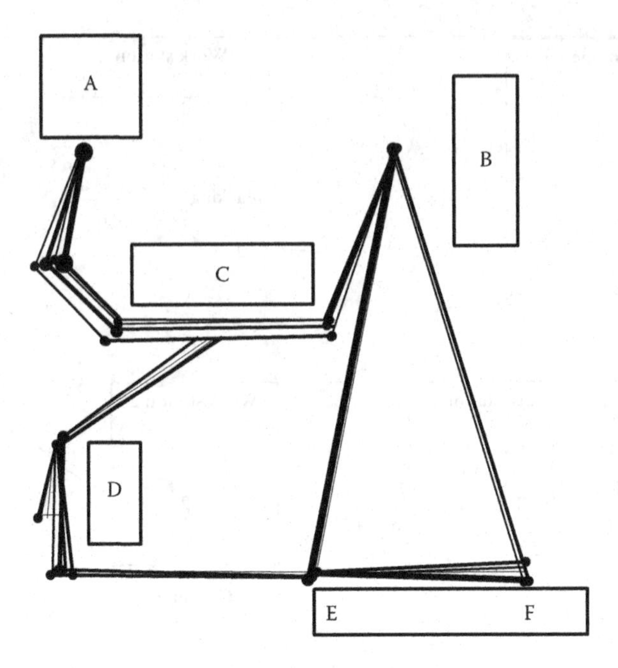

Figure 2.11 Typical string diagram.

to another within a working area. Thus, a string diagram is a special type of flow diagram.

The observation period must be a prolonged one as the movements' sequence changes from one cycle to another. It is also possible to record the movements on a video prior to transferring them on to a string diagram. The following procedure may be used to prepare a string diagram:

- A preliminary observation to determine the places (points) visited by the worker.
- Preparation of a scale model of the layout area where the places visited (machines, workstations, etc.) lie.
- Assigning a code (number or letter) to each place visited by the workers. The code could be simple (single letters) or more complex (series of two or three letters, numbers, or alphanumeric). A complex code is useful for instance when different types of materials are handled or when different locations at the same work station are used.
- Record in writing each movement of the operator. If the movements are long, arrival and departure times may also be recorded. A simple form tabulating the times and from–to movements may be used for recording movements (e.g., machine A to inspection station B—11; time elapsed for these moves, if necessary, may be expressed as an average of the range given).
- Once the observations are complete, pins or nails, are inserted in all labeled locations on the scale model. Pins/nails are also inserted in locations where obstructions are present (e.g., machine corners, pillars, etc.).
- A large length of string (thread) is measured (in feet, yards, or meters) and is wound around the pins/nails in the same order as the worker movements have been recorded.

- The length of the remaining (unused) string is measured and subtracted from the original length to determine the total distance covered by the worker.

Figure 2.11 shows a typical string diagram. The paths that are traversed more frequently have greater number of strings. These are the paths (movements) that should be examined first.

2.2.10 Cyclegraphs and Chronocyclegraphs

Both these diagrams require a light source (e.g., LEDs) and record the path of the movement on a photographic film. The camera shutter remains open throughout the recording. Both these diagrams are quite accurate but unsuitable for introductory work.

While the cyclegraph uses a continuous light source, the chronocyclegraph uses an interrupted light source. In a cyclegraph, the path of the movement appears as a continuous line while the chronocyclegraph records the movement path as a series of dots (these appear pear shaped on photographs). The interval between dots provides information about limb acceleration and deceleration (large accelerations and decelerations are undesirable, as they reflect the use of larger force exertions and, therefore, greater metabolic energy expenditure). A light source is attached to the hands (or the joint of interest) and the movement is recorded as the hands move during the course of the movement.

This methodology is quite common in kinematic and kinetic biomechanical analysis. In such analysis, lights are attached to several joints (wrists, elbows, shoulders, etc.) and movements are recorded stereoscopically.

2.2.11 Travel Charts

String diagrams are very effective, but they take a long time to construct. These are also difficult and cumbersome to construct when the movement patterns are complex. In such situations, travel charts are helpful. These charts provide a tabular presentation of information about moves (movements) of workers, materials, and equipment between a number of departments (places, workstations, or points).

A travel chart is also known as *from–to Chart* as it really provides information about the number of moves from each location (department) of interest to other locations (departments) of interest. The travel charts are always square and have departments listed as rows and columns. The columns indicate "movement to" and the rows indicate "movement from" alternatively, columns can be listed as "movement from" and rows as "movement to." Each square (cell) in the chart represents a department or place. If, for instance, movements between 10 different departments are to be charted, the chart will have 10 rows and 10 columns leading to 100 squares. A diagonal line is drawn across the top left corner to the bottom right corner. This diagonal line crosses squares representing movement from the same department to the same department. The squares in the first row represent the departments from where the movements begin. The squares in the first column represent the departments to which the movements are made.

To record the movement from one department to another, the observer enters the column representing the department from where the move takes place and moves down along that column until the row representing the department, where the move ends, is reached. An indication, in the form of a mark, is entered in the square representing the square at the intersection of the column and row of interest. All moves that take place over a period of interest are recorded in this manner. Once all the moves are recorded, the number tally (total) for each intersecting square is put in that square.

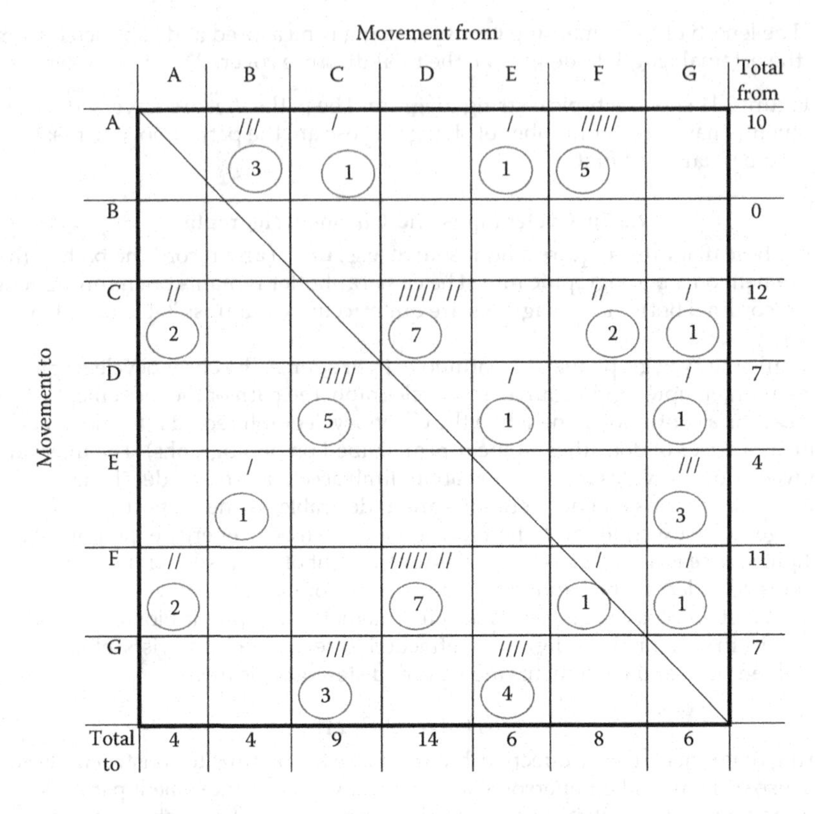

Movement to \ Movement from	A	B	C	D	E	F	G	Total from
A		/// 3	/ 1		/ 1	///// 5		10
B								0
C	// 2			///// // 7		// 2	/ 1	12
D			///// 5		/ 1		/ 1	7
E		/ 1					/// 3	4
F	// 2			///// // 7		/ 1	/ 1	11
G			/// 3		//// 4			7
Total to	4	4	9	14	6	8	6	

Figure 2.12 Typical travel chart showing the number of movements between seven departments.

The purpose of travel charts is to identify departments that have the heaviest traffic so that these could be placed next to each other in order to minimize the total distance traveled and the associated costs (cost of energy, time, effort, etc.). Figure 2.12 shows a typical travel chart. The column totals show all the movements from each department and the row totals show all the movements to a department. The information provided in the travel chart, thus, can be used to improve the workplace layout.

2.3 EXAMINING THE RECORDED INFORMATION

Once the information regarding activities is recorded, it needs to be subjected to a critical and objective examination. Three techniques are discussed in this section:

- The Ishikawa or fishbone diagrams
- The questioning technique
- The operations analysis

The Ishikawa diagram technique has limited application but nevertheless it can prove useful as it can identify causes that result in a certain effect. Once the causes are identified, remedies can be developed and the negative effect can be mitigated.

The questioning technique, on the other hand is a very powerful tool that allows thorough examination of activities. This tool is widely used in identifying causes of extra work content and finding solutions to eliminate or minimize them.

The operations analysis requires asking questions regarding 10 different but interdependent factors on a specially designed form. Answers to several different factors may be sought at the same time—it is not necessary to complete answering questions pertaining to a single factor before moving on to another.

2.3.1 Ishikawa or Fishbone Diagrams

Named after its developer, Japanese Professor Ishikawa, these diagrams are now just referred to as the fishbone diagrams and are very popular in the Japanese quality circles. These diagrams help in identifying customer needs and wants and converting them into design and manufacturing requirements through the quality function deployment (QFD). The fishbone diagrams help in identifying "causes" to an "effect"; the effect is the "head" of the fish and "causes" are the bones. The goal is to identify as many bones (main causes) and minor bones (subcauses) as possible. Solutions that eliminate these bones also lead to the elimination of the effect. The fishbone diagram is prepared by first identifying the effect (fish head) of a specific problem.

The body of the fish (bones indicating causes) is then added to the head. The goal is to add as many major and minor bones as possible (it is desirable to have at least three levels of bones; back bone, major bones, and minor bones). The diagram creates an overview of the factors (causes) that contribute to the problem (effect).

Different strategies may be used to define major bones. Bones may be developed along the factors that contribute to productivity (humans, equipment, materials, methods, type of energy, policy, facilities, etc.). Figure 2.13

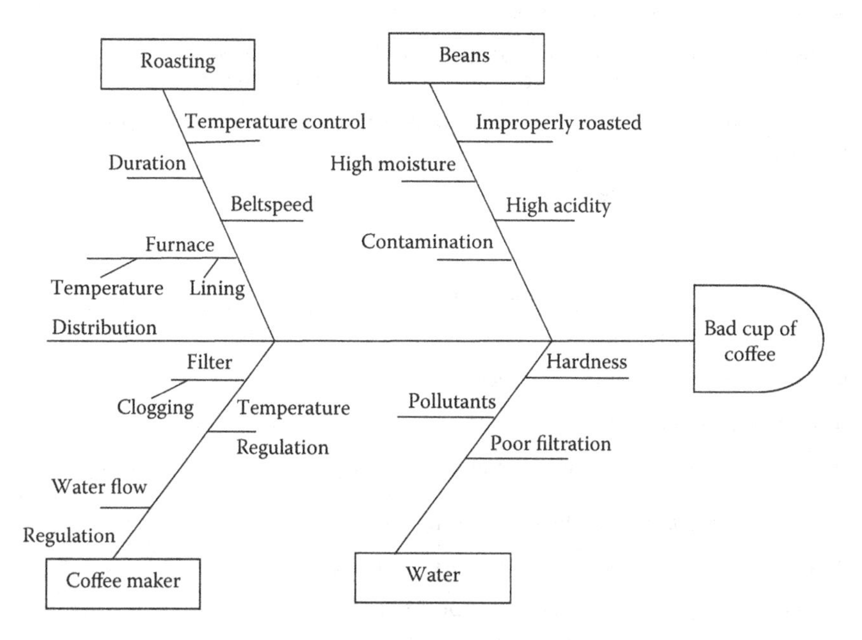

Figure 2.13 Fishbone diagram for a "bad cup of coffee."

shows an elementary fishbone diagram for the effect "bad cup of coffee." The causes, bones, can be several. In certain cases, minor bones to the fourth level are shown. The diagram is not intended to depict all possible causes (exhaustive), but shows how the cause–effect relationship, particularly causes, may be developed.

2.3.2 The Questioning Technique

The questioning technique provides a very powerful means by which recorded information is subjected to critical examination. Each recorded activity is subjected to a series of progressive questions, systematically. The activities that have been recorded fall under two categories:

- Activities in which the object (material or work piece) is being modified in some manner (e.g., its shape is being changed by machining) or is being examined or is being moved
- Activities that include object awaiting some action (delay) or is in storage

The first category includes activities that are

- Preparatory in nature (e.g., loading, setup, etc.)
- Operational in nature (e.g., machining, assembly, etc.)
- Postoperational activities (e.g., unloading, inspection, etc.)

While the operational activities add value to an object or material, the preparatory and postoperational activities do not. It can also be said that since operational activities add value to materials or objects, these are productive activities and all others (preparatory and postoperational) are nonproductive (adding to the excess work content). The overall purpose of the questioning technique is to

- Eliminate preparatory and postoperational activities.
- Reduce operational activities (if such activities are reduced, many preparatory and postoperational activities are automatically eliminated).

The questioning technique requires that examining questions be divided into two sets:

- A set of **primary questions**
- A set of **secondary questions**

The set of primary questions aims at

- Eliminating operational activities
- Combining operational activities
- Simplifying operational activities
- Reordering or rearranging operational activities

These are accomplished by questioning the

- **Purpose** for which such activities are performed
- **Place** at which these activities are performed
- **Sequence** in which these activities are performed

- **Person** who performs these activities
- **Means** by which these activities are performed

These lead to a set of *primary questions*:

Purpose *What is actually being done?*
Why is this activity really necessary?

These questions aim at eliminating the operational activity.

Place *Where is this activity being done?*
Why is it being done at this place?

These questions aim at combining operational activities.

Sequence *When is the activity performed?*
Why is it being performed at that time?

These questions aim at increasing the effectiveness of operations by rearranging the sequence of activities.

Person *Who is performing the activity?*
Why is this particular person performing the activity?

These questions aim at making the outcome more effective.

Means *How is the current activity being performed?*
Why is it being performed in this manner?

These questions are intended to simplify the activity.

Once these set of questions are asked, it is ascertained what is being done, why it is being done, where it is being done, who is doing it, and how it is being done. The next step in the questioning technique is to subject the answers to primary questions to a second set of questions. This second set is called the *secondary questions*. The set of secondary questions intends to focus on "what else," "what should," "who else," "who should," and so on.

The following set combines the primary and secondary questions:

Purpose *What is being done?*
Why is it being done?
What else could (might) be done?
What should be done?

Place *Where is it being done?*
Why is it being done there?
Where else might/could it be done?
Where should it be done?

Sequence *When is it being done?*
Why is it being done then?
When might/could it be done?
When should it be done?

Person *Who is doing the work?*
Why is that person doing it?
Who else might/could do it?
Who should do this work?

Means *How is it being done?*
Why is it being done that way?
How else might/could it be done?
How should this work be done?

The success of the methods study depends upon asking and successfully and completely answering these questions every time the recorded data are analyzed. Once the answers are recorded, improvement in methods can be undertaken.

2.3.3 Operation Analysis

An operation analysis is somewhat similar to the questioning technique but the kinds of questions that are asked are little different and pertain to more specific factors. It is suggested that at least the following nine factors be considered when analyzing a job or activity and relevant questions be asked pertaining to each (The list of questions provided is suggestive, not comprehensive.):

- Purpose of the operation
  - Are all operations necessary? Is the purpose achieved?
  - Can the vendor perform the operation?
  - Can the operation be eliminated? Combined?
  - Can the results be achieved in the previous or following operations?
- Design of the part
  - Is the part necessary? Are all components required?
  - Can standard off-the-shelf parts be used?
  - Is the design cost-effective?
  - Are all design features necessary?
- Process analysis
  - Is the sequence of operation the best possible?
  - Can the operation be eliminated? Combined? Performed elsewhere?
  - Can some other process be used?
  - Is the process cost-effective? Can it be automated?
- Inspection requirements
  - Are tolerances specified really necessary? Purpose of these tolerances?
  - What kind of quality control process should be used? Is it really necessary? Is the process in control?
  - Is inspection effective?
- Materials
  - Is the material used most suitable?
  - Can a cheaper material be used?
  - Can the amount used be reduced? By reducing part size? By reducing part volume? By reducing part weight?
  - Can a more expensive material reduce processing requirements? Number of operations?
  - Can the material allow proper and economical packing?

- Materials handling
    - Are the devices being used appropriate? Right size containers and bins? Proper cranes and conveyors?
    - Are conveyors taking full advantage of gravity?
    - Has backtracking been avoided?
    - Is contamination during materials handling avoided?
    - Has the distance between the starting and ending points minimized?
- Workplace layout
    - Has the layout been evaluated for tools, storage, movements, etc.?
    - Are the tools supported?
    - Is there a place for everything?
    - Is everything in place it is supposed to be?
- Working conditions
    - Is the light adequate?
    - Is the noise exposure within acceptable levels?
    - Have safety issues been addressed?
    - Are there adequate provisions for hydration, rest, etc.?
- Method
    - Have the principles of workstation design used in evaluating methods? (See Section 2.4 for details)
    - Are methods cost-effective?
    - Have alternatives been considered?

As in the case of the questioning technique, detailed and careful answers to the questions raised above would lead to minimizing the excess work content.

In this section, we have described three different techniques to examine the recorded information. Use of any one technique does not preclude using the other two techniques. In fact, it is recommended that all the three techniques be used to find ways to eliminate excess or unproductive work.

2.4 MAKING IMPROVEMENTS

At the end of a critical examination, answers are provided to the questions that have been raised. These answers lead to improvements in methods and elimination or minimization of excess work. The improvements are reflected in a number of ways. Following is a list of likely improvements in methods/activities (The list is neither exhaustive nor in any order of importance.):

- Reduction in the number of operations (by elimination or combination)
- Reduction or elimination of inspection
- Elimination of backtracking
- Reduction in materials handling (reduction in the number of pickups and putdowns)

- Increased cost-effectiveness
- Reduction in delays
- Reduction in idle time
- Increased utilization
- Reduction in the distance traveled (a straight line movement between the starting and ending points is optimum)
- Smoother materials flow
- Reduction in material cost
- Increased process efficiency
- Efficient scheduling and production control
- Enhanced safety (reduced number of injuries)
- Better workplace layout
- Better facilities layout
- Reduced fatigue
- Better quality
- Reduced work in process
- Increased use of standard parts
- Parts' consolidation
- Improved working conditions
- Increased productivity

This is just a partial list of improvements that can be expected from a proper and detailed methods study.

One of the major places to make improvements is at the workstation or the workplace. Since in a plant there are numerous workstations, the room for making improvements there is very significant. Keeping this in mind, *workstation design* is addressed in a separate section (Section 2.5).

2.5 WORKSTATION DESIGN

In this section, we address issues dealing with one worker working at a workstation or workplace. The attention in designing workstations is on the worker. The overall goal is to increase worker productivity through reduction of worker fatigue, increased worker comfort, increased efficiency of worker movements, provision of job and workplace aids, etc. A number of factors that help to achieve these are discussed in this section.

2.5.1 Static Work/Loads and Fixed Working Postures

Static work or static load, sometimes also referred to as isometric load (due to prolonged muscular contraction; muscle length does not change), is bad for the worker as it reduces the supply of energy to the muscles as well as removal of waste products from the muscles. Both energy supply and waste removal are accomplished through the supply of blood that is curtailed during static loads. Using fixed work postures, such as standing in a place or sitting for long periods, is a form of static loading.

Static workloads and fixed postures lead to the following physical problems:

- Increased blood pressure (it is higher for standing than for sitting)
- Increased limb pressure (e.g., on feet when standing for long periods without walking)
- Increased cardiovascular effects (cardiac output, heart rate, etc.)
- Increased physical fatigue
- Swelling of limbs (e.g., feet swelling during stationary deskwork; also noticed during long flights)
- Increased intervertebral disk pressure if the spine is curved

Static workloads and fixed work postures, therefore, should be avoided. The following solutions aid in alleviating the problems of static work and fixed postures:

- Design the workplace layout for alternate sitting/standing posture
- Make the activity rhythmic (e.g., shifting body weight from side to side)
- Ensure that shoes are cushioned, provide a large area of contact with the floor (thus reducing pressure), support ankles, slip resistant, and have metallic cover for toes (for safety). Shoes should be "earthen" (flat heels instead of high heels)
- Make the floors soft or provide mats so that the pressure is reduced and there is resiliency
- Provide support for feet in the form of bar rails
- Support limbs (arms, head, neck, trunk, etc.) to prevent localized muscle fatigue
- Provide back support; the support should allow the torso to move forward and backward as desired by the worker

2.5.2 Repetitive Trauma Disorders

Repetitive trauma disorders also known as cumulative trauma disorders or repetitive strain disorders are the result of

- Frequency of the activity
- Extreme joint movement
- Force

Weaker workers show the symptoms of this syndrome sooner than stronger workers. Eventually, however, even the stronger workers succumb and their bodies fail. The efforts should therefore be to minimize cumulative or repetitive trauma disorders.

The most prevalent of the repetitive traumas in industry is *carpal tunnel syndrome (CTS)* (it should be noted that while many workers suffer from CTS, the majority of CTS are not occupationally related and may be the result of recreational activities or are idiopathic).

CTS is injury of the median nerve of the wrist and results in a decrease in the effective cross section of the carpal tunnel. The median nerve swells from repeated contact with the hard surface of the bone or tendon or ligament.

The occurrence of CTS results in pain, wasting of muscles at the base of the thumb, weaker exertions, and clumsiness. The following approaches are recommended to reduce work-related incidences of CTS:

- CTS is more influenced by repetition than force. Therefore, the number of cycles per wrist should be reduced
- Consider automation or mechanization
- Consider job enlargement (add nonrepetitive activities to the job in place of repetitive ones)
- Consider job rotation (to nonrepetitive jobs)
- Avoid repeated exposure to vibrations between 40 and 130 Hz
- Reduce the joint angle by changing the job posture (by changing worker or object orientation)
- Reduce the magnitude of force exertions
- Reduce the duration of force exertions
- Use motorized tools
- Use sharp knives
- Use power grip (whole hand) instead of precision grip (finger grip)
- Use trigger strips instead of trigger buttons
- Use tools that have grips made out of resilient materials
- Make grips larger
- Use springs to open tools such as scissors and clippers
- Use gloves that have a better fit and no seams on the palmer side
- Support tools so the operator does not have to (hanging them from the ceiling or a support at the top; tool is pulled down for work and retracts back up when done)

2.5.3 Work Height

Work height is defined by the elbow height. It would be different if the worker is sitting or standing; it is not a fixed height from the floor. Work height is important in the design of workstations. If it is too high, shoulders must be raised. This could lead to painful cramps. If, on the other hand, the working height is too low, then the back would be bent excessively causing back pain. Since people vary in stature, it is critical that the work height be adjustable. Adjustability may be provided by allowing work to be raised or lowered. Alternatively, the platform on which the worker is sitting/standing could be raised or lowered. Allowing the work to be raised or lowered is easier than making the platform height adjustable (sometimes it is easier to provide elbow height adjustability by providing adjustable height chairs).

It needs to be kept in mind that work may require the use of force (use of the upper body) or may require dexterity (use of hands and fingers). If the work requires manipulation, that is the use of fingers and hands, the work height should be 50 mm below the elbow height. This height is slightly below the heart level and leads to higher productivity and lower physiological costs (lower heart rate and metabolic energy expenditure). Work that requires manipulation and precision must be raised above the elbow height for improved vision. Again, work height adjustability is the key.

Work that requires use of force should be 150–200 mm below the elbow. This allows the worker to use the upper arm and shoulder muscles to make the forceful exertions. Again, repeated exertion of force is fatiguing and, therefore, work height adjustability is necessary. Figure 2.14 shows the concept of work height adjustability for different kinds of work.

The work height recommendations are summarized below:

- Provide an adjustable work height by allowing the surface supporting work to be raised or lowered
- For work that requires use of fingers and hands, the optimum work height is 50 mm below the elbow height
- For work requiring precision, it needs to be above elbows; the elbows and arms must be supported
- Work requiring force should be 150–200 mm below the elbow height
- Provide an adjustable height chair

2.5.4 Providing an Adjustable Chair

In the standing posture, the surface area of the shoes supports the entire body weight. This puts heavy pressure on the feet. If the worker sits down, almost 80% of the body weight gets support thus reducing pressure on the feet significantly. This also leads to fatigue reduction and reduced metabolic energy expenditure. The benefits of a chair are

- Weight is taken off the legs
- Unnatural body postures, such as standing on one leg, are avoided
- Metabolic energy expenditure is reduced (lower oxygen consumption)

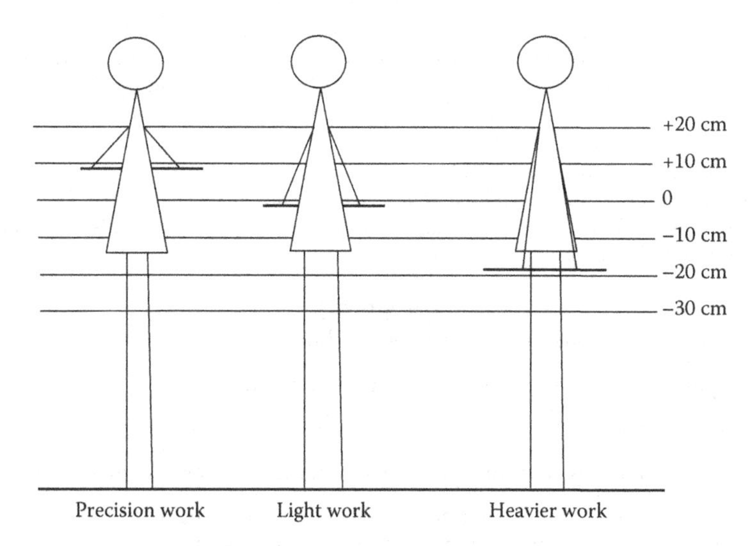

Figure 2.14 Adjustable work heights for different kinds of standing work.

- Demands on the circulatory system are reduced
- Fewer complaints of backache

There are certain disadvantages as well. Prolonged sitting leads to weakening of the abdominal muscles, causing digestive problems, and adversely affects the curvature of the spine. Therefore, the workstation should be designed to permit alternative sitting and standing postures. The cost of a chair is significantly lower than the cost of lower productivity resulting from the inability to sit.

The following are the design recommendations for a good chair:

- Seat should allow height adjustability between 38 and 53 cm.
- Seat breadth should be between 40 and 48 cm.
- Seat depth should be between 33 and 47 cm.
- Seat should have a "waterfall" front with an angle between 4° and 6°.
- Seat should have a nonslip breathable surface (fabric is desirable if there is no seat ventilation, as in cars).
- Backrest should be adjustable between 95° and 105°.
- Back rest should be movable in a vertical direction and should have a height between 10 and 23 cm; breadth between 30 and 40 cm; horizontal radius between 31 and 46 cm; and vertical radius (shape) should be convex.
- Backrest surface should have low friction.
- Armrests should have dimensions of: 15–28 cm length, 4–9 cm breadth, and 16–25 cm height.
- Chair should have five legs with casters and foot support.

2.5.5 Feet/Legs and Hands

Workstation design should allow the use of both hands and feet. While the feet are slow, they have greater muscle mass (compared with hands) and allow greater force/power generation. Greater muscle mass also allows prolonged (continuous) force exertion without legs getting tired. The feet, however, do not have the dexterity and flexibility of the hands. Hands are flexible and allow greater ability to manipulate and control.

From a workstation design standpoint, both feet and hands should be used: feet for activities requiring force/power (for instance in operating pedals) and hands for manipulation and precision work.

2.5.6 Gravity

Gravity is free potential energy. It can be used to

- Feed parts (e.g., vibratory feeders and gravity chutes)
- Dispose parts
- Transport materials from one place to another (e.g., roller conveyors)

Opposing gravity, on the other hand, costs energy with upward movement. This is true even if the movement is manual. Depending on the size and weight of the work, bigger limbs may be needed (fingers, hands, arms, etc.). *Therefore, vertical upward movements are not desirable and should be avoided; all movements should preferably be in the horizontal plane.*

2.5.7 Momentum

Both time and energy are required to increase speed (acceleration) or to slow down (deceleration). This happens when movements in the workplace change directions suddenly—movement must come to rest (deceleration), change direction, and then accelerate in another direction. Such motions are strenuous and expensive in terms of time and energy as greater muscular effort is required. Such motions also adversely affect performance. *Therefore, circular motions are desirable (directional change is gradual and motions must not come to a sudden stop and then accelerate).*

2.5.8 Principles of Motion Economy and Hand Motions

Human movements at the workplace should be of the lowest possible order. The following is a classification of the order of movements:

- Class 1 pivot—knuckle finger (body member)
- Class 2 pivot—wrist fingers and hands (body members)
- Class 3 pivot—elbow forearm, hands, and fingers (body members)
- Class 4 pivot—shoulder upper arm, forearm, hands, and fingers (body members)
- Class 5 pivot—trunk torso, upper arm, forearm, hands, and fingers (body members)

As the order of movement increases, from Class 1 to Class 5, more limbs get involved. This increases physical strain, and time and energy requirements of completing the movement. Therefore, it is desirable to use the lowest order of movements at the workplace.

The work and workplace should be designed such that, whenever possible, the movements of both hands should begin and finish at the same time. Further, the movements should be symmetrical and in opposite direction. The exception is when movements require eye–hand coordination; in such cases, movements should be parallel. Two-handed motions should also be rowing motions as opposed to alternating motions, as alternating motions require more force (10% more) and cause higher heart rates.

Two-handed motions are more efficient—it takes longer but cost/hand is lower. This applies to cost/unit as well as metabolic cost.

Movements pivoted about the elbow are fastest, more accurate, and least strenuous and therefore workstation movements should be pivoted about the elbow (only forearm movements). The exception is cross-body movements that are more accurate when pivoted about the shoulder. Moreover, free-swinging movements are faster, easier, and more accurate than controlled (restricted) movements.

2.5.9 Dominant Hand

The dominant hand (preferred hand) is 10% faster and 7% stronger than the other hand. It is also more accurate as it gets more practice. The majority of the population, approximately 90%, is right handed (this means that 1 out of 10 is left handed). The dominant hand also has greater reach and better grasp. *Therefore, work should enter the workstation from the side of the preferred hand and leave from the side of the nonpreferred hand.*

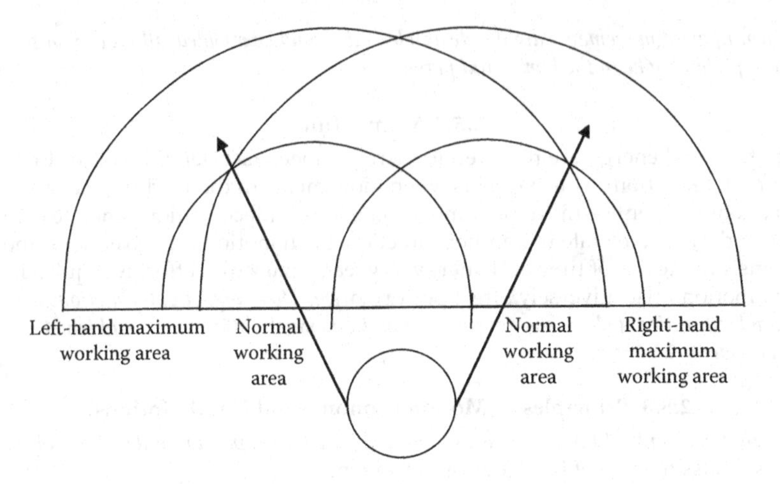

Figure 2.15 Normal and maximum working areas.

2.5.10 Working Area

Figure 2.15 shows the normal and maximum working area in the horizontal plane. The normal working area is the reach area in the horizontal plane when movements are pivoted about the elbow. The maximum working area is the reach area in the horizontal plane when movements are pivoted about the shoulder. *Work should be placed in the normal working area. The maximum working area should be used as a last resort and for placing containers for supplies.*

The normal and maximum working areas decline as the height from the horizontal plane increases or decreases. Figure 2.16 shows the maximum reach area for the average U.S. men and women in sitting and standing postures.

If the movement requires eye–hand coordination, work should be placed within the area of intersection between the "visual comfort zone" (area between two semicircles 3"and 15" in radius) and the normal working area.

2.5.11 Physical "Fit"

People come in various sizes. The workstation design should accommodate majority of workers. A shorter worker has smaller reach while a larger worker will have bigger dimensions. Although both smaller and larger individuals need to be accommodated, the cost of including "everyone" is prohibitively expensive and is not always technically possible.

To include a large proportion of the population, we use the following design rules:

- Design for one extreme (this rule would include everyone)
- Design for a specific population by eliminating people in the lower and upper percentiles (e.g., middle 95%)
- Design for the average (the 50th percentile; inconvenience everyone)

The last design rule is rarely used as it does not accommodate anyone. It is typically used in applications such as the design of the park bench or the height of a checkout counter in a grocery store.

The first design rule is often preferred, if applicable, because it accommodates everyone. For instance, designing a door that allows the tallest person to

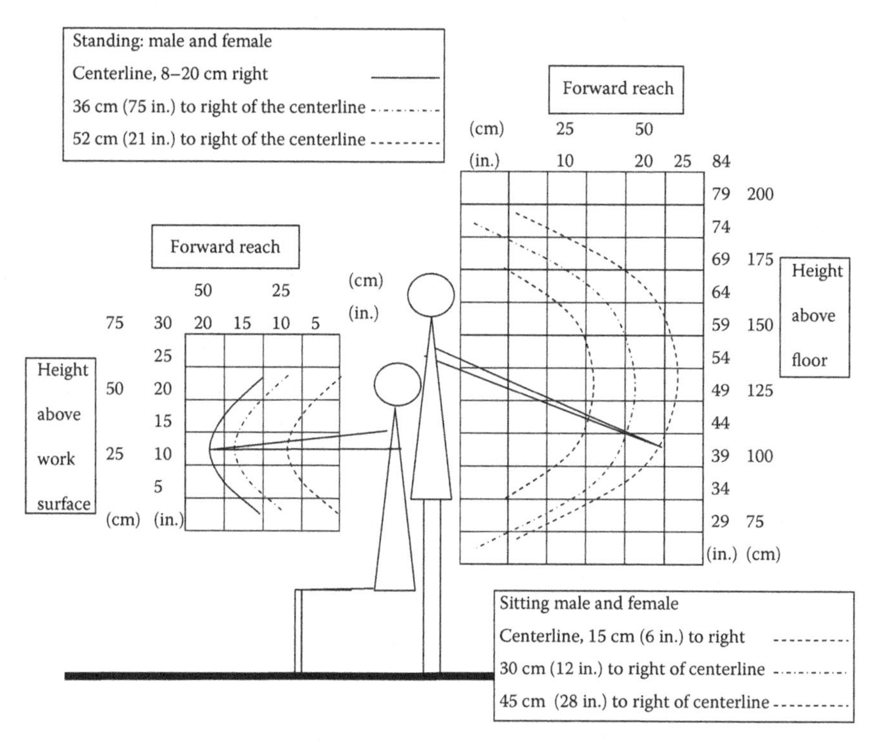

Figure 2.16 Reach distances for average U.S. male and female populations.

walk through (upper extreme) also allows everyone else who is shorter to walk through as well.

Generally, we end up using the middle design rule—design for a specific population. We select a population, such as the middle 90% or 95%, and provide adjustability within that population. For example, chairs that can be designed to adjust to body size of the middle 95% of the population. The workstation needs to be designed so that the largest person in the specified population can fit and the smallest person in that population may reach all areas.

2.6 SUMMARY

This chapter summarizes the basics of methods study which aims at minimizing idle, or excess, work. A number of techniques are described for recording activities for critical examination. A variety of charts and diagrams for recording activity-related information have been described. This is followed by the description of techniques that allow objective examination of the recorded information; three distinct techniques are described for this purpose. Besides describing the form of resulting improvements, we have provided guidelines for designing workstations.

SUMMARY

3 The Basics of Time Study

Time study is the second component of work study. It follows the methods study, described in Chapter 2. Time study is defined as a technique to measure and record the time it takes by an operator to perform a specified task under specified conditions at a specified rate.

The various advantages of time study, or work measurement, have been outlined in Section 1.4 and range from labor cost estimation to establishing production capacity; all critical to enhancing productivity. Before the time study is carried out it is important to ensure that methods study has been undertaken, all excess work has been eliminated, and the total work content of the job is as close to the basic work content (minimum work required to perform the task) as possible.

3.1 SELECTING THE JOB TO BE TIMED

A job may be selected for time study for a number of reasons including the following ones:

- No time standard exists for the job
- The job is a new one
- Methods study has just been performed on the job and the job has changed
- The job appears to be "expensive" as it is taking too much time
- There is a dispute between the union (or the worker) and the management as to how long the job should take
- The equipment utility at the workstation is low and needs to be improved
- A new design, method, equipment, or tool has been introduced
- The work performed is causing delay downstream (station starvation) owing to large time it is taking; this also may be causing work pile up (bottlenecking) upstream
- The work standard has been in place too long and needs to be revalidated
- An incentive system needs to be implemented

As stated above, time study should not be conducted until the methods study has already been carried out (exceptions would be for reasons such as dispute). Until the work method has been defined and prescribed (standardized), the work content would vary and so would be the time it takes to perform it; the work standard applies to a specific way of doing things (method) and if the job method changes, the standard would no longer apply to that particular job.

One of the standard problems that arises in selecting the job for timing is found within companies that pay on the basis of piece rate. If it was discovered that incentive payments to a worker were excessive and could not be maintained, then it would appear that the work standard at that workstation was too copious and needed to be reassessed. Such a move, more than likely, to be opposed by the worker (or the union), as any attempt to establish an objective work standard, would result in lower payments. Here the job selection is not based on job's importance based on Pareto Distribution, but for an alternative reason. In these circumstances, it is better to study other jobs until the persons conducting the time study have established their reputation for fairness and objectivity.

Since establishing work standards is directly tied to "expected output" or "performance" and may determine a worker's earnings, it is important that the worker, the supervisor, and the workers' representative (or union representative) be apprised of the purpose of the study upfront. The worker that is selected for the time study should be "qualified." A qualified worker is an average worker who is suitably motivated and has adequate experience and skills to perform the job to acceptable levels of quality and quantity in a safe manner (selecting above average or below average worker is not desirable as it is difficult to assess their pace. Further, selecting an average worker improves the overall acceptability of the work standard by workers at a higher rate). Often, this is a contentious issue particularly when workers' representative pushes for a "representative" worker as opposed to a "qualified" one. Under such circumstances, it is important for the objectivity of the time study to push for an average worker who performs at a steady rate and who is not bothered being observed while working. This is also reflected in fairly low variations in cycle time. A large variation in cycle time may reflect that: (1) the worker is trying on purpose to increase the cycle time, (2) the worker is not qualified, or (3) there are factors such as variations in materials, tools, equipment, or method used by the worker. These factors must be corrected prior to continuing with the time study. It should be kept in mind that the reasons for large variations are relatively easy to detect for repetitive jobs but not for jobs that require a lot of skill and judgment on the part of the worker. Under such circumstances, it is critical that the method be precisely established prior to timing it. Besides, variations in materials, equipment, and tools must be looked into prior to conducting the time study.

The worker who has been selected for the time study should be apprised of the purpose of the study. *Timing an operator without his/her knowing or from a hidden location must never be done.* It is not only unethical but also gives the impression that there is something to hide. For developing an honest and objective work standard cooperation and trust of the worker are paramount. Timing operations without workers' knowledge is no way to build trust.

In the event there are several qualified workers, it may be desirable to perform the time study on several of them.

3.2 RECORDING THE INFORMATION

Figure 3.1 shows a very basic and generic form for collecting time study data. This form can be customized by adding information as needed. At the very least, a time study form should show the following information:

- Information about the facility, workstation, equipment, and tool (if applicable)
- Information about the worker (name, number, title, etc.)
- Information about the person conducting the time study (name, title, etc.)
- Information about the work being timed (process, materials, drawing number, part number, operation number, etc.)
- Information about the date and time the study was conducted, including start and finish time
- Any special inspection and quality requirements
- Name and title of the person checking the information
- A sketch of the workstation layout (may be put on the back of the form)

Time study summary sheet		
Department:	Plant:	Study No.:
Operation:	Process:	Sheet No.: of
Plant/Machine:	Material:	Start time:
Tools and gauges:	Light:	End time:
	Noise:	Elapsed time:
Product/Part:	No.:	Worker:
DWG No.:	Inspection:	Stop watch No.:
Quality:	Working conditions:	Studied by:
Worker:		Date:
		Checked by:

Cycle No.	El. No.	Element description	R	WR	ST	OT	Element description	R	WR	ST	OT
Note: R = rating, WR = watch reading, ST= subtracted time, and OT= observed time.											

*Workstation layout on the back.

Figure 3.1 Typical time study sheet for continuous time recording.

- Listing of all the elements comprising the job
- Rating (pace) for each element
- Recorded times (stop watch readings, subtracted times, and observed times)
- Information about continuation sheets, if used (continuation sheets may or may not show all the information included on the first sheet. For instance, it is not necessary to show information about the process or the operator on every sheet used during the study)

Since multiple cycles are timed during a time study, each element is timed multiple times and continuation sheets become necessary. Hence, it is necessary to use a summary sheet (Figure 3.2). The normal time for each element in the summary sheet is the average time for that element, recorded over several cycles and given in the time study sheet (total time for an element/frequency of occurrence of that element during the study).

Time study summary sheet		
Department:	Plant:	Study No.:
Operation:	Process:	Sheet No.: of
Plant/Machine:	Material:	Date:
Tools and gauges:		Start time:
Product part:	No.:	End time:
DWG No.:	Inspection:	Elapsed time:
Quality:	Working conditions:	Studied by:
Worker:	Stop watch No.:	Checked by:
Sketch and nouns on back of sheet 1.		

El. No.	Element description	NT	F	#Obs.	

Note: NT: normal time, F: frequency/cycle, and #Obs: total number of observations.

Figure 3.2 Typical time study summary sheet.

It is important to keep in mind that all the information in the time study form should come from direct observation. Further, all relevant information must be recorded.

3.2.1 Breaking the Job into Elements

As evident from Figure 3.1, a description of the elements making up the job is required for timing. To complete a job, a sequence of elements must be performed. For the purpose of time study, an element is defined as a part of a job that is distinct and can be easily timed (measured). For instance, picking up a workpiece from the bin, loading it on the machine, and performing the drilling operation would be an element. Holding the workpiece, cleaning it, unloading it, and putting it in another bin would be another element.

It should be noted that not all elements may appear in each cycle that is observed for timing. In general, there are eight distinct types of elements:

- Repetitive elements—these are elements that occur in every work cycle
- Occasional elements—these are elements that do not occur in every cycle but may occur at regular or irregular intervals (e.g., inspecting the machine setting every fifth cycle)
- Constant elements—these are elements for which the observed time remains unchanged (e.g., removing and inserting cutting tools in the machine)
- Variable elements—these are elements for which the observed time varies (e.g., pushing cart to different departments)
- Manual elements—these are elements performed by the worker
- Machine elements—these are elements performed by the machine
- Governing or principal elements—these are elements that have the longest time within a work cycle than any other element being performed concurrently (For instance, using go-no go gage periodically while turning a shaft to a specified diameter.)
- Foreign elements—these are elements that are extrinsic to the job (e.g., wiping nose after sneezing)

There are certain guidelines that help break down a job into elements. The following may be helpful:

- Elements should be as small/short as can be timed easily
- Each element should have a definite beginning and end that can be easily identified
- Elements should be distinct from each other and should form a single natural motion (e.g., reaching, grasping, moving back, loading, positioning, and clamping)

Additionally

- Manual and machine elements should be separated
- Elements should be studied over a number of cycles before being timed
- Constant, variable, and occasional elements should be separated from each other
- Occasional elements should be timed separately

Once the job has been broken down into elements, we must determine how many times each element must be timed. That is, the determination of the sample size or the number of observations or the number of readings for each element. For us to do so, we need to perform a preliminary time study and use statistics.

Let us say that a preliminary time study has been conducted and an element has been timed N' number of times. The value of each reading is X_i, where i takes a value from 1 to N'. From this, we can calculate the number of times this element needs to be observed, N, to be

$$N = \left(40\left(\sqrt{N' \, \Sigma X^2 - (\Sigma X)^2}\right) \Big/ \Sigma X\right)^2$$

where Σ represents the summation of all X or X^2 values. This equation provides a ±5% margin of error or relative accuracy. The required number of observations, N, will need to be increased if one desires greater accuracy (or lower margin of error). An increase in the number of observations, N, will also increase the cost (time). A lower value of N will consequently also reduce the accuracy but lower the cost. For example, for 10% accuracy the value of N will be

$$N = \left(20\left(\sqrt{N' \, \Sigma X^2 - (\Sigma X)^2}\right) \Big/ \Sigma X\right)^2$$

To demonstrate how to use this equation to calculate the number of observations (sample size) for an element in a time study, let us say that a preliminary study has observed the following times for an element:

X	X^2	
6	36	
5	25	
7	49	
6	36	
$\Sigma X = 24$	$\Sigma X^2 = 146$	$N' = 4$

Putting the values for ΣX, ΣX^2, and N' in the above equation for a ±5% error, we have

$$N = \left(\frac{40\left(\sqrt{4(146) - (24)^2}\right)}{24}\right)^2$$

or $N = 22.27$ or 23 readings. This means that an additional 19 observations would be needed to get an absolute accuracy of 5%. As more observations are taken, N should be recalculated; perhaps with a total of 10 observations or 15 observations. As more data accumulate it may be observed that the sample size up to that point is adequate ($N = 10$ or 15). On the other hand, the required value of N may increase beyond 23.

3.2.2 Timing the Elements

There are several methods that are available and used to record elemental time. These are discussed in detail in Chapter 10. In this section, we only discuss the timing procedures that utilize a stopwatch (mechanical or electronic). A variation of the stopwatch procedure using multiple stopwatches is also described. In all cases, the elemental time is recorded live (direct observation by the time study specialist).

Primarily, there are two methods for timing the elements using a stop watch:

- Continuous or cumulative timing
- Snapback or flyback timing

In the *continuous timing method*, a single stopwatch that runs continuously throughout the study is used. The stopwatch is started at the beginning of the study and it is allowed to run until the end of the study. At the end of each element, the stopwatch time is recorded. The time may be recorded in the following graduated scales:

- One-fifth of a second; one revolution of the stopwatch is equal to 1 min
- One-hundreds of a minute; one revolution of the stopwatch is equal to 1 min (this type of stopwatch is also called a decimal-minute stopwatch and can record up to 30 min)
- One-hundreds of an hour per revolution

The decimal-minute stopwatch is the most widely used mechanical stopwatch. It is started and stopped by a sliding button on the side of the watch. This type of watch can be used for continuous recording as well as snapback recording. The observers, however, must have their line of sight aligned with the stopwatch as well as the element being performed. At the end of the element, the elemental time, indicated by the watch hand, is recorded while the stopwatch continues to record time. Thus, the time elapsed until the completion of each element is recorded. The time for each element is determined by subtracting from the elapsed time until the end of an element the elapsed time until the end of the previous element. For instance, if the cumulative time at the end of element 1 is 0.3 min and the cumulative time at the end of element 2 is 1.2 min, the time for element 2 is obtained by subtracting 0.3 min from 1.2 min giving a time of 0.9 min. When the study is completed, the total of the time for each element is added and it must be equal to the total elapsed time (ending time minus starting time). This is a check on the accuracy of recording.

The most noticeable advantage of continuous time method is that since the watch is never stopped, any activity that is missed or omitted is not excluded from the overall cycle time. The continuous time method is preferred by workers and strongly favored by the trade unions in the United States for this very reason.

In the *snapback* or *flyback* method, the stopwatch hand is snapped back at the end of the element. The watch starts recording the next element while the observer records the time for the previous element. Since each elemental time is recorded individually, there is no need for subtraction as in the continuous time method and a pattern of elemental times can be discerned. As in the case of continuous time method, the observer's line of sight must be aligned with the element being performed and the stopwatch.

The snapback timing method however has several disadvantages:

- It is difficult to read times when short elements are performed sequentially
- It is difficult to read moving watch hands and to remember where it was
- The observed time is shortened slightly as it takes time, even though short, for the watch hand to come to zero and start again
- There is a possibility that the observer may record a foreign element or irregular elements as delays and stop the watch instead of continued recording
- When confused, the observer may simply stop the watch
- There is a lack of "check" as in the case of continuous timing method, making it difficult for workers to accept

A *three-watch system* alleviates the problems mentioned above. The system has three stopwatches that are all controlled by the same lever. The hand of the first stopwatch starts moving initially, the hands of the second watch are stopped at some value, and the hands of the third stopwatch are at zero. At the end of the first element, the observer pushes the lever and the first watch hand moves to zero, the second watch starts recording, and the third watch hand stops at a time value. The time for element 1 is recorded from the hands of stopwatch 3. At the end of element 2, the observer pushes the lever again and stopwatch 2 stops at a time value, stopwatch 1 starts recording, and stopwatch 3 hands go to zero. The time for element 2 is recorded from stopwatch 2. This process continues until all elements are timed. The three-watch system thus has all the advantages of a single stopwatch continuous recording and the single stopwatch snapback recording.

Electronic watches are widely used. The digital display provides an accurate recording of times as compared with analog watches as there is no moving hand; when the observer pushes the button, the display freezes, allowing time to be recorded, while the clock continues to record time. These days electronic watches are used in conjunction with a laptop computer or a computing device built into the *time study boards*. This makes recording elemental time much simpler.

When a job has short cycle comprising short elements, *differential timing method* may be used. In the differential timing method, time for several sequential elements is recorded with and without an element. The time for the element is then determined by subtraction. For instance, time for elements 1–4 is first recorded and then time for elements 1–3. Subtracting the second time from the first provides time for element 4. Elements can be grouped in different ways to provide time for each element.

3.2.3 Rating

Once a job has been selected for time study, it is broken down into elements which are then timed. The next step in the process of establishing work standards is to assess the worker's rate of working (the one who is being observed) as compared with the observer's (person conducting the time study) concept of the standard rate (pace) of working. Since this concept is subjective, rating is a controversial aspect of setting time standards. This also makes setting time standards process an inexact one. If time standards are used for setting up incentive pay, the process can be contentious.

Before discussing rating, it is important to revisit the concept of the "qualified" worker and the "average" worker. As said earlier, the worker who should be observed performing during the time study should be a qualified worker and, if possible, a number of qualified workers should be studied prior to formulating the time (work) standard. A qualified worker neither performs very slowly nor very fast. Who is a "qualified" worker? A qualified worker is one who has "adequate experience" and is "suitably skilled" to perform the specified work in a manner that is satisfactory from a quality, quantity, and safety standpoint. How do we determine what are adequate experience and suitable skill levels? The following help us to determine the answers for such questions:

- The worker should be able to achieve consistent and smooth motions
- The work should have a "rhythmic" quality, as opposed to a strained one
- The worker should have a short reaction time to signals or stimuli
- The worker should be able to anticipate issues and problems arising in the course of performing the work and should be able to overcome them in short order

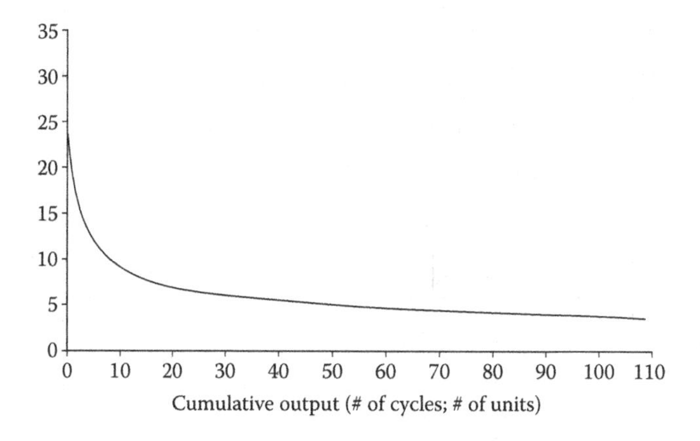

Figure 3.3 Typical learning curve.

- The worker's performance should be relaxed, without an appearance of self-consciousness
- The worker should be able to demonstrate that he possesses the skills the job requires and not appear to be harried or fumbling

Since various tasks require a variety of skills, from physical strength to mental concentration to visual acuity to specialized knowledge, not all workers may necessarily be suited for a time study. It may take a certain amount of time to acquire the skills necessary to perform a task. Without reaching that degree of proficiency, the worker may not achieve the "qualified" status. One way to determine if an appropriate level of proficiency is reached is to look at a worker's performance or by learning progress curve. Figure 3.3 shows a typical progress curve, with cost on the y-axis and cumulative output on the x-axis. The curve follows the equation

$$Y = AX^B$$

where A and B are constants; Y is the cost; and X is the cumulative output.

The above equation can also be plotted as a straight using a log–log scale as shown in Figure 3.4. The reduction in cost (indicated by %) every time the cumulative output doubles is used to indicate what percentage curve the learning curve is. For instance, if the cost is reduced by 10% when the cumulative output increases from 4000 to 8000, it is called a "90% curve"; that is the cost at 8000 units is 90% of the cost at 4000 units. A typical learning in the manufacturing industry varies from 68% to 98.5%. By simply plotting the cost/unit with the cumulative output on a log–log graph paper the learning rate can be obtained. As Figure 3.3 shows, after a certain number of units, there is no appreciable decline in cost. This is also indicated by the horizontal part of the curve. This indicates that the learning is over and the worker has reached the "qualified" status.

The rate of learning (performance improvement) depends on how much can be learned. The amount that can be learned depends on the background of the worker and technology (better technology improves learning). As an organization improves its learning (through better designs, machinery, and technology), individual learning improves as well.

While we can define a "qualified" worker in objective terms (e.g., on a learning curve), defining an "average" worker is simply not possible as there is no such

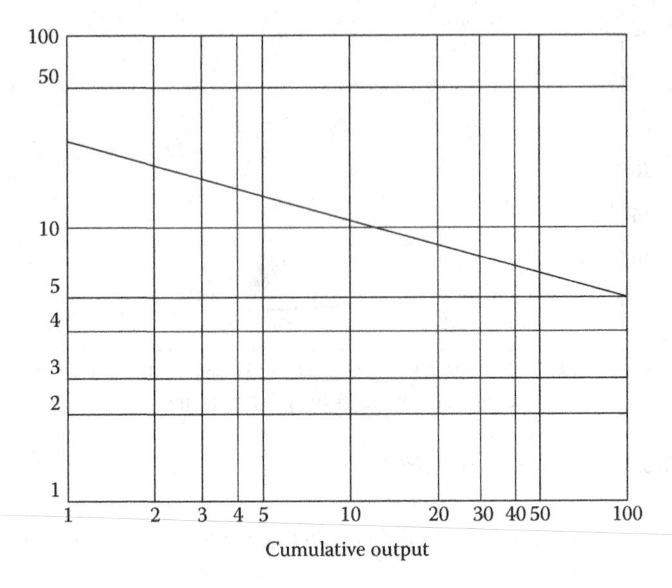

Figure 3.4 Learning curve using a log-scale.

worker who can be described as an "average worker"—it is merely an idea or a statistical myth. For example, if half of the population are men and half of the population are women, what is the average gender?

Nevertheless, we use the concept of average when we plot a particular worker attribute graphically, say height, weight, or performance time. In such cases, a pattern emerges. This pattern in case of most human attributes resembles a "normal distribution." This normal distribution provides an average, and workers who fall within the middle 32.8% of the distribution may be considered average as far as their performance proficiency is concerned. However, developing the distribution of performance times in order to select an "average" worker who can be timed is not realistic. Imagine determining performance times of 100 workers (or more) and plotting them to obtain the average profile. The time study person, therefore, must rely on a "qualified" worker.

For a time standard to have any relevance, it must be achievable as far as the majority of the workers are concerned. If it is too high (too little time to do the job), only a small percentage could achieve them and estimates (e.g., expected output) would never be attained. On the other hand, if the time standard is too low (too much time to do the job), everyone would easily achieve them including the slowest worker and it would be inefficient and economically unaffordable. Therefore, the time standard must be fair. How is this accomplished? The time study person must be able to compare the pace at which the worker being observed is working with some standard level. This standard level of pace is defined as the pace a qualified worker would achieve naturally and without overexertion over the entire working day. This standard pace is also called the *motivated productivity level (MPL)*. The MPL indicates the human work capacity and is the maximum level at which the worker may be expected to work. This MPL is modified by a term called *"expectancy"* to achieve a pace or level of performance called the *acceptable productivity level (APL)*. The APL is a work pace that is generally considered satisfactory to both the management and workers (unions) and is also known as the *normal productivity level or normal pace*:

Normal or Acceptable Productivity Level = MPL – Expectancy

Thus expectancy is really a political compromise. Once the definition of APL is agreed upon, the time study person can rate the performance of the worker being observed compared with the defined standard (APL). The APL is also called a performance with a rating of 100%. Walking at a speed of 3 miles per hour or dealing a deck of 52 cards in 30 s are paces with an APL, or rating, of 100%.

To review, the purpose of rating is for the time study person to determine from the time taken by the worker being observed the time it would take a qualified worker (physically fit, suitably motivated, adequately experienced, and skilled) to perform the job. If in the time study person's opinion, the qualified worker would take less time than the worker being observed, a rating of less than 100% is assigned. On the other hand, if it is thought that the qualified worker would take more time to complete the job then a rating greater than 100% is assigned to the job. In both cases, it is the pace of operation being performed by the worker being observed that is being rated. Operations that require judgment (e.g., inspection) are harder to rate and the time study person must be experienced in such tasks to rate them.

There are two methods that are widely used to rate job performance pace:

- Pace rating method
- Objective rating method

The pace rating system is the simplest system and requires rating the speed or tempo. It is also the most widely used rating system. The time study person concentrates on motions such as reach and move, while avoiding stationary motions, to rate the job performance pace.

The objective rating method uses three steps as follows:

- Step 1—rate the pace of the job
- Step 2—rate the difficulty of the job (going uphill is more difficult than going downhill)
- Step 3—multiply the speed factor by difficulty factor to get the overall factor

In rating jobs, it is important that

- The time study person has some experience with the job (all persons conducting time study must be thoroughly trained and must gain experience. This is particularly important if the jobs being timed require judgment, movement of heavy loads, has lots of stationary work elements or require different set of skills of varying degrees)
- It is better to rate each element at least twice and then average the ratings
- For short elements, elements may be grouped by category
- Since pace varies from one cycle to another, adequate number of cycles should be studied and ratings averaged

It should be noted that there are many factors that influence the pace of performance, and thus rating, that are beyond the control of the worker. For instance

- The quality of the material
- Wear and tear of tools over the life of such tools, affecting working efficiency
- Changes in environmental conditions
- Different elements requiring different levels of attention, etc.

If a sufficient number of cycles are studied, these problems can be alleviated. *There are a number of rating scales but the scale that rates APL as 100% is recommended.* On this scale, a rating of 50% would mean "very slow; clumsy, fumbling movements; worker inattentive or lacking interest." A 125% rating would mean "very fast; worker exhibiting a high degree of assurance, dexterity, and coordination of movements that are well above the average worker." A 150% rating would mean "exceptionally fast; effort unlikely to be kept up for long durations as intense effort and concentrations are required; pace levels that can be achieved by only a very few."

3.2.4 Allowances

Despite increased mechanization and automation, numerous jobs still require effort from a human element. With that being said, allowances must be made for recovery from fatigue. Allowances also must be made for personal needs, such as visits to the restroom and replenishing lost body fluids. In addition, allowances may be needed for delays that are beyond the control of the worker. All these allowances are generally categorized as personal (P), fatigue (F), and delay (D) (PDF) allowances. These allowances are discussed in detail in the next chapter, Chapter 4. A brief discussion of allowances, however, follows in this section.

Just as rating the pace of worker performance is contentious, so is the determination of allowances. Together, these two aspects of the time standard determination process make time standards subjective. It is known that standards for the same task vary across organizations due to varying levels of APL and permitted allowances. Moreover, it is difficult to determine allowances that apply precisely to all workers under all circumstances for the following reasons:

- Workers vary in physical conditioning, age, experience, etc. As a result, different people require different amounts of fatigue allowances in order to recover.
- Variations in work, from light to heavy, lead to different requirements for recovery from fatigue. Working around hot processes, such as around furnaces, for example, is very different from working on an assembly line.
- Environmental conditions, such as those caused by heat, humidity, noise, dust, etc., affect buildup of fatigue differently and different allowances would be needed for relaxation and recovery from fatigue.
- Personal needs also vary. As a result different allowances are needed for personal needs.

Given the variations at the personal level, work level, and environmental level, it is easy to see why it is not possible to develop any standards for the determination of allowances.

Of the three types of allowances described above (PDF), personal and fatigue allowances require that time be added to the normal or basic time (average observed time × rating). Delay allowances are added only if conditions warrant such an addition. Together, personal and fatigue allowances are also referred to as relaxation allowances. Obviously, relaxation allowances depend on the nature of job and environmental conditions. These allowances allow workers to recover from physical and mental fatigue (weariness) and restore the body's vitality.

Relaxation allowances are generally added as a percentage to the normal time for each element separately. The resulting times for all elements, when added, yield the standard time for the job. Adding relaxation allowances for each element separately also permits accounting for severe working conditions, such as the radiant heat from a furnace, that could exist for some elements but not all. If elements do not vary much from each other, relaxation allowances can be calculated by applying it to the normal time for the whole job.

> Allowances for climatic conditions that may result from working outdoors are applied to the entire shift (working day) rather than to the element or the job thus allowing the time standard for the job to remain independent of climatic changes caused by seasonal changes.

In the United States, it is common to have three prescribed breaks per 8-h shift—two 15-min breaks (e.g., one in mid-morning and one in the mid-afternoon) and a 30-min lunch break at mid-day. Relaxation allowances are generally part of these breaks. If the total relaxation allowance time exceeds these prescribed breaks, it may be taken at the discretion of the worker. *However, if the working conditions are severe (excessive heat, cold, noise, etc.), management should make rest pauses mandatory. A proper work–rest schedule must be implemented to ensure that workers do not suffer from any adverse health effects. Further, these mandatory rest pauses should have proper rest places and facilities for cold and hot drinks and snacks.*

Unlike personal and fatigue allowances (relaxation allowances), delay allowances are not worker dependent. The delay allowances are beyond the control of the operator as they reflect unavoidable delays resulting from issues such as machine or tool breakdown, machine maintenance, receiving instructions from the supervisor, etc. In cases like these, when such situations cannot be avoided, a small allowance is added to the normal time (the delay allowance must not exceed 5% of the normal time).

In addition to delay allowances, special allowances are occasionally added as necessary. Examples of such special allowances are

- Machine or tool cleaning allowance
- Start-up allowance (e.g., heating up the die prior to operations)
- Shutdown allowance
- Setup or changeover allowance

Such special allowances are generally given as a part of the working day (shift time) and do not form a part of the time standard.

3.3 DETERMINING THE TIME STANDARD

Now that all the necessary information has been recorded, it is possible to set the time standard for the job. The average observed time obtained from the summary time sheet shown in Figure 3.2 is simply called the *observed time.* This observed time needs to be adjusted to reflect the performance time for a qualified worker by applying the rating. The adjusted time is called the *normal time* (also called the basic time).

$$\text{Normal time} = (\text{Observed time})\,(\text{rating})$$

To this normal time, allowances (PDF) are added. The resulting time is called *standard time*. Allowances can be added in two ways:

- As a percentage of shift time (time at work)
- As a percentage of work time (actual time spent working)

If the allowances are given as a percentage of shift time, the standard time is given as follows:

$$\text{Standard time} = \frac{(\text{Normal time})}{(1 - \text{Shift allowances})}$$

If allowances are given as a percentage of work time, the standard time is given as

$$\text{Standard time} = (\text{Normal time})\,(1 + \text{Work time allowances})$$

In some countries, such as the United Kingdom, it is required that the standard time be calculated by giving allowances as a percentage of work time instead of the shift time. In such cases, the second formulation for standard time would apply. *It should be noted that if the same percentage allowance is given either as a shift allowance or as a work allowance, giving it as a shift allowance is more advantageous to the workers.* This can be easily illustrated using a simple example. If the shift allowance is 5% and the shift duration is 8 h (480 min), the working duration would be 456 min (=480 – 0.05 × 480). For the same amount of working time, 456 min, the work time allowances will have to be 5.26% (=480/456).

3.4 DETERMINING STANDARD TIME WHEN WORKING WITH MACHINES

There are situations when a worker has to work with a machine or operate multiple machines. In such cases, determining standard time is slightly more involved. Let us take the case of one worker and one machine first. We are presuming that the method study has already been carried out and the worker–machine cycle time has been minimized. In a one worker–one machine scenario, three different situations can possibly occur:

- Machine is working and the worker is not working (waiting on the machine to finish)
- Machine is working and the worker is able to do other things in the mean time
- Worker is working and the machine is idle (waiting for the worker to finish)

The time that the machine is working is called *machine-controlled time*. The time when the worker is working but the machine is not is called *outside work time*. The time when both the worker and the machine are working is called the *inside work time*. Figure 3.5 shows these various times for a hypothetical situation.

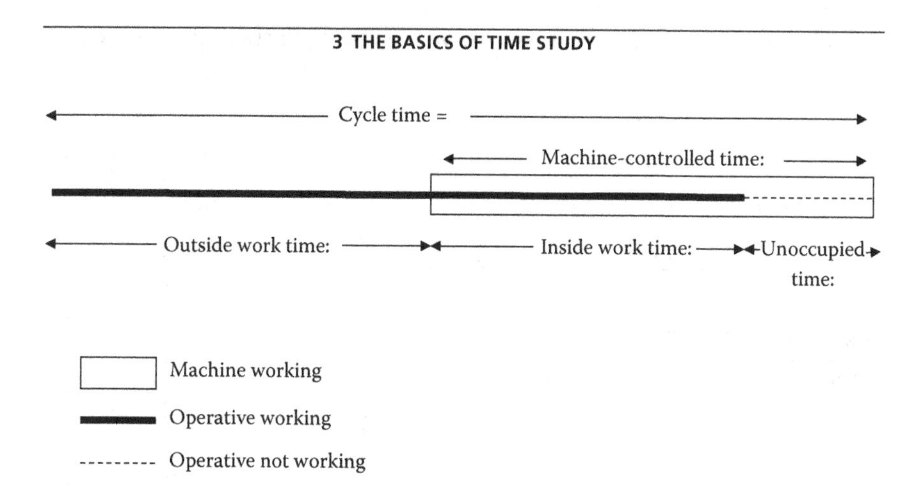

Figure 3.5 Cycle time for a hypothetical single worker–single machine system.

In these situations, the observed time is determined on the basis of the cycle time. The calculation of relaxation allowances, however, is different. First, the relaxation allowance needs to be separated into personal allowance and fatigue allowance. The personal allowance must be based on the entire cycle time which includes the machine-controlled time and the outside work time as the entire time is spent at the workstation. Calculation of fatigue allowance and its application to the time standard, however, is a little more complicated. The fatigue allowance is calculated on the basis of the time (of the cycle) that is actually spent working. This means the time of the cycle the worker remains unoccupied is not used in calculating the fatigue allowance. Once the fatigue allowance is determined, the time study person needs to determine how much of this allowance can the worker be expected to take during the unoccupied time. The balance can then be added to the cycle time. It may also be possible for the worker to take some of the personal allowance during the unoccupied time if the cycle time is long.

There are circumstances when a worker has to operate multiple machines and the work is cyclic in nature. For instance, situations involving screw-making machines where the worker primarily performs loading and unloading machines as the machines are semiautomatic in nature. In such situations, standard times may be calculated as has been described above.

Situations involving multiple workers and multiple machines are considered beyond the scope of this book. All one can suggest is that the unoccupied time must be evaluated by conducting extended studies.

3.5 A TIME STUDY EXAMPLE

A simple job from a nonmetal processing industry is used here as an example to demonstrate the application of time study. Such cyclic jobs are typical in many industries. The following is a description of the job.

A major producer of athletic goods receives unwashed, knitted material that must be dyed and processed, then folded into bundles by a machine, called a calender. These bundles are 15 in. (38.1 cm) wide and weigh about 45 lbs (20.45 kg). The job requires a worker to remove each bundle from the calender which is about 53 in. (135 cm) from the floor to a table 25 in. (63.5 cm) high and tie it. The

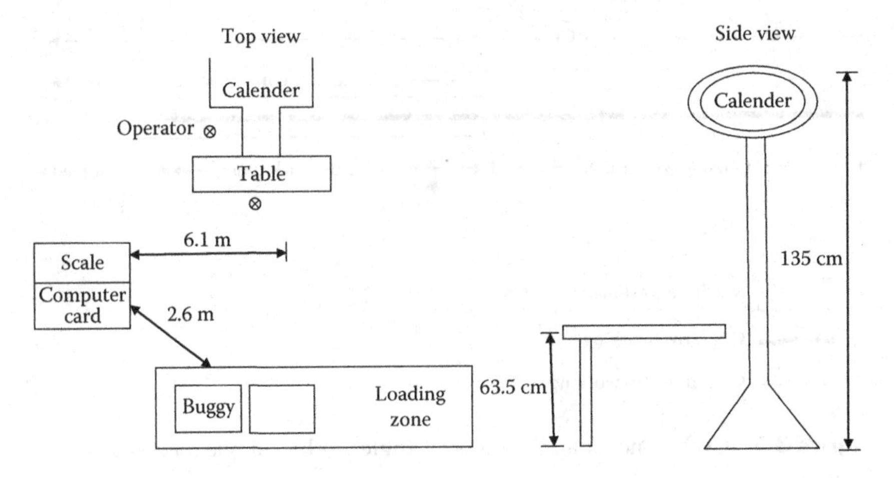

Figure 3.6 Schematic of the workplace.

bundle then has to be carried a distance of 20 feet (6.1 m) at knee height to a scale, weighed, and tagged with a computer generated card. The bundle then has to be loaded on a buggy for temporary storage. The buggy is 12 feet (2.6 m) away from the scale. Bundles are trucked from the buggy to another plant for sewing. Figure 3.6 shows the schematic of the workplace.

The plant operates 7 and 1/2 h each day. This job has undergone methods study and needs to be timed so that a time standard can be established. The worker selected for observation is a male "qualified" worker, 33 years of age, who is physically fit and has performed this type of job for 3 years at this plant.

The job has been broken down into five elements. Table 3.1 provides a description of these elements.

A three-watch system was used to conduct the time study and five cycles were timed initially to determine how many cycles should be studied. On the basis of the values of this initial sample and cycle times (not elemental times), it was determined that nine cycles should be timed in order to have an accuracy of ±5%.

Figure 3.7 shows the elemental times observed (watch readings) for the first three cycles, along with the ratings, for demonstration purposes; times are in minute. Observed times were directly obtained from the three-watch system as there was no need for subtraction. The summary sheet with the averaged normal time for each element is shown in Figure 3.8.

Table 3.1: Description of Elements for the Example Job

Element	Description
1	Lower the bundle from the calender to the table
2	Tie the bundle
3	Lift the bundle
4	Carry the bundle to scale, weigh it, and tag it
5	Carry the bundle to the buggy for temporary storage

Time study sheet		
Department: 11	Plant: Alabama	Study No.: 1
Operation: Bundle tagging	Process:	Sheet No.: 1 of 3
Plant/Machine: Calender	Material:	Start time: 8:30 a.m.
Tools and gauges:	Light:	End time: 1 p.m.
	Noise	Elapse time: 4.5 h
Product/Part:	No.:	Worker: M 12
DWG No.:	Inspection:	Stop watch No.: 17 T-W
Quality:	Working conditions: Normal	Studied by: MCK
		Date: 7/4/11
		Checked by: CKG

Cycle #	EL.#	Element description	R	WR	ST	OT	Element description	R	WR	ST	OT
1	1	1	105	-	-	0.209					
		2	100	-	-	0.186					
		3	100	-	-	0.058					
		4	100	-	-	0.310					
		5	105	-	-	0.193					
	2	1	110	-	-	0.210					
		2	85	-	-	0.180					
		3	100	-	-	0.06					
		4	105	-	-	0.30					
		5	100	-	-	0.20					
	3	1	105	-	-	0.211					
		2	95	-	-	0.188					
		3	100	-	-	0.059					
		4	105	-	-	0.306					
		5	110	-	-	0.191					
	4	.									
		.									
		.									
		.									

Note: R= rating, WR= watch reading, ST= subtracted time, and OT= observed time.
*Workstation layout on the back.

Figure 3.7 Elemental times for the example.

The job described in this example is a physical job, performed standing up, and involves materials handling over a prolonged period of time. Buildup of physical fatigue is a major issue in this job. This is in addition to time required for personal needs. Since all elements of the job are somewhat similar (materials handling type requiring physical effort), allowances were added to the normal cycle time which was 0.962 min. A personal shift allowance of 5% and a fatigue shift allowance of 7% were added to the normal time (see Chapters 4 and 9 for details regarding the determination of allowance percentages). The resultant standard time is calculated as follows:

$$\text{Standard time} = \frac{\text{Normal time}}{(1 - \text{Shift allowances})} = \frac{0.962}{(1 - 0.12)}$$
$$= 1.093 \text{ min/cycle}$$

Time study summary sheet					
Department: 11	Plant: Alabama				Study No.: 1 Sheet No.: 1 of 1
Operation: Bundle tagging Plant/Machine: Tools and gauges:	Process: Material:				Date: 7/6/11 Start time: End time:
Product part: DWG No.: Quality:	No.: Inspection: Working conditions: Normal				Elapse time: Studied by: MCK
Worker: M 12	Stop watch no: 17 T-W				Checked by: CKG
Sketch and nouns on back of sheet 1.					
El. No.	Element description	T	F	Obs.	
1	Lower the bundle	0.210	9		
2	Tie the bundle	0.187	9		
3	Light the bundle	0.059	9		
4	Carry the bundle to scale, weigh it, and tag it	0.310	9		
5	Carry the bundle to the buggy for the temporary storage	0.196	9		

Note: T: time, F: frequency/cycle, and #Obs: total number of observations.

Figure 3.8 Summary sheet for the example.

This means the expected output per day would be handling of 411 bundles (=450 min/1.093 min). Since the actual cycle time is 0.962 min, the accumulated time difference per cycle (=53.84 min; 0.131 min/cycle × 411 cycles) is taken as relaxation allowances. The company allows the worker to stagger this rest period as he needs.

4 Allowances

As mentioned in Chapter 3, there are two factors that introduce subjectivity, and hence inaccuracies, in the determination of time standards. The first one is the determination of rating and the second is the application of allowances. While one can reduce the subjectivity from rating by rating each elemental performance at least twice and training the time study operatives, careless application of allowances makes it difficult to reduce the subjectivity it introduces into time standards. The determination of allowances for personal needs and unavoidable delay are straightforward, yet the application of allowances for fatigue is often problematic. In this chapter, we first discuss personal and delay allowances and then fatigue allowances.

As also mentioned in Chapter 3, rest allowances are added to the normal time either as a percentage of shift time or as a percentage of actual work time. *It is recommended that allowances be added to the normal time as a percentage of shift time unless specified otherwise. Doing so is favorable to the workers.*

4.1 PERSONAL ALLOWANCES

Every worker requires personal allowances to take care of personal needs such as

- Going to the restroom
- Having a drink of water
- Wiping the face
- Keeping the workplace clean and tidy, etc.

There are no scientific bases for determining the allowance percentage for personal needs. The published literature indicates personal allowance variation from 5% to 10%. In addition, different personal allowance percentages have been recommended for men and women (this is no longer legal in the United States). *It is now widely accepted that a 5% allowance for personal needs is sufficient for general working conditions.*

In the United States, each shift typically has a 30-min meal break. Many organizations also have a 10–15-min break in each half of the shift. Although a worker can use these breaks for personal needs, most organizations do not consider these as a part of personal allowances.

4.2 DELAY ALLOWANCES

There are two kinds of delays possible: (1) avoidable delay and (2) unavoidable delay. No allowances should be provided for avoidable delays as, obviously, such delays are avoidable. Unavoidable delays, on the other hand, need to be compensated as these are beyond the control of the worker. Examples of unavoidable delays are

- Machine breaking down
- Materials not arriving on time
- Bottlenecks upstream resulting in station starvation downstream and worker waiting for work to arrive
- Supervisor providing instructions to the worker
- Setup needs
- Cleaning and lubricating the equipment

- Work delays due to handling of multiple machines (machine(s) may have to wait for operator attention. This is also known as machine interference)
- Catastrophic tool failure, etc.

In case the delay exceeds 30 min, worker may be required to clock out to work on something else. Workers could also use some personal time during prolonged delays.

Providing delay allowances, which could vary from 0% to 10%, compensates for prolonged delays. The most popular method of determining delay allowances is work or occurrence sampling. This technique is discussed in detail in Chapter 5.

In general, unavoidable delays go down over months and years as the organization becomes more attuned to the causes, such as material-related problems (delivery and quality). This requires that the time standard be redetermined; otherwise, it would become too loose (unnecessary and unneeded allowances remain).

A special case that results in unavoidable delays is handling of multiple machines by single or multiple operators (one operator and multiple machines or several operators and multiple machines). This generally happens with automatic or semiautomatic machines and the operator is mostly involved in loading and unloading operations. In this case, a machine may have to wait for the operator to load it or unload it, resulting in the overall delay of work and extended work cycle. As the number of machines grows, the situation complexity increases. While there are ways to determine delays in such situations (refer to the suggested reading list at the end of this book), dealing with it here is considered beyond the scope of this book.

4.3 FATIGUE ALLOWANCES

The term *fatigue* is nebulous in nature, pertaining to the human in both body and mind. The Webster dictionary defines fatigue as: "The exhaustion of physical and mental strength." This qualitative definition of fatigue offers no insight into the origin, mechanisms, manifestations, or processes involved with the development of fatigue. The published literature offers three working categories of fatigue:

- Subjective fatigue
- Objective fatigue
- Physiological fatigue

Subjective fatigue is characterized by decrements in psychological factors such as mental alertness, the ability to concentrate, and motivation. It is believed to be the result of reduced levels of activity of the reticular activating system. Objective fatigue is characterized by decrements in the output of work, while physiological fatigue is characterized by identifiable changes in the physiological processes such as depletion of energy reserves or lack of energy supplies. In general, fatigue denotes loss of efficiency and disinclination for any kind of effort. It is not a single or definite state that can be clearly defined. If a distinction is made between muscular fatigue and general fatigue, the term becomes less ambiguous. This is because the forms of fatigue, muscular and general (mental, visual, auditory, etc.), result from completely different physiological processes: muscular fatigue is the result of depletion of energy reserves or due to lack of energy supply, while general fatigue (feeling of weariness) is the result of reduced levels of activity of the reticular activating system.

The general effects of fatigue are

- Reduced output/unit
- Reduced efficiency and productivity
- Increased discomfort
- Weaker muscular contractions and depleted energy resources
- Increased levels of physiological responses (heart rate, metabolic energy expenditure rate—oxygen consumption, and body core temperature)
- Loss of concentration
- Nervous fatigue (from visual and auditory exposures)
- Increased risk for overexertion and injury

If there was no fatigue, then there would be no need for fatigue allowances. There is no task that is not fatiguing in some manner. Even sitting on a sofa for long periods of time can be fatiguing—fatigue in this case may result from maintaining the same posture for prolonged periods. Thus, there is a need to compensate workers for the time lost due to fatigue. Therefore, the need for fatigue allowances.

Fatigue allowances have two distinct parts: (1) the constant or basic part and (2) the variable part. Historically, the management and workers across many industries have agreed, as reported by the International Labor Organization (ILO), that a 4% allowance added to the normal time would take care of the fatigue needs of a worker performing light work in a seated position under acceptable working conditions. This allowance is known as the *basic fatigue allowance*. If the task or the working conditions additionally tax workers' physical or cognitive systems, additional fatigue allowances known as *variable fatigue allowances* must be added to the basic fatigue allowance of 4% (It should be noted that the total of all allowances is used to adjust the normal time in order to determine the time standard).

Sometimes event fixed multipliers are used to determine fatigue allowances. For example, in the case of methods time measurement (MTM) predetermined time system (see Chapter 6 for details regarding predetermined time systems) it is recommended that 8% of the MTM time be added to the MTM time, for rest allowances, for all jobs regardless of the work, the worker, or the environment. In such cases, the allowances become inherently inaccurate since individual and situational differences are ignored. The resulting work standards, consequently, are unreliable and tend to vary from one industry to another. For example, the workers normally perform under some sort of incentive scheme, where their performance is in the 120%–130% range, as compared with the 80%–90% range without incentive payment. In such cases, the time standards developed through conventional industrial engineering practice have no relationship with the physiological stress involved in moderate-to-strenuous work. Empirically determined fatigue allowances may also be either inadequate or unnecessary. If the fatigue allowances are inadequate, the very purpose of providing such allowances is defeated and the time/unit increases. However, if the fatigue allowances are excessive, the result is waste and suboptimum performance. As we approach the limits of human productivity, in terms of performance, it becomes critical that individual variability be considered. The assumption that there are no individual differences, as fixed multipliers in the case of MTM would indicate, is very dubious indeed.

As mentioned earlier, one of the effects of fatigue is decline in output. This decline has been suggested as a means to determine fatigue allowances very early on, in the twentieth century. Using the decline in output as a measure, fatigue percentage allowance for continuous work is calculated as

$$F = \left\{ \frac{(T - t)}{T} \right\} \times 100$$

where T is the operational time at the end of work and t is the operational time at the beginning of work.

The widely used fatigue allowances come mainly from three works published in the 1970s, and differ widely. Each of these sources reports the fatigue rest allowances under three categories:

- Physiological fatigue allowances
- Psychological fatigue allowances
- Environmental fatigue allowances

Physiological fatigue allowances take into consideration the weight, force, or pressure, and sometimes the worker gender. Psychological fatigue allowances consider the effect of visual and mental workload and its consequent strain. Environmental fatigue allowances consider the effects of temperature, humidity, noise, illumination, and supply of air.

Not only do the recommendations from these three sources differ widely from one another, but they also consider different factors. Table 4.1 shows the recommendations from the three sources. (*Individual sources are not identified, as identifying them is not considered relevant. For those interested in these sources refer to "allowances topic" in titles listed under suggested reading.*)

The allowances recommended in Table 4.1 are obtained on the basis of the sum of minimum and maximum allowances for individual factors in their respective category. The values are a percentage of the total time. Table 4.1 shows the significant differences between the recommendations made by different sources in the case of maximum allowances in each category.

Besides wide differences in allowance ranges, the suggested use of these values also differs. While source 1 and source 3 values are to be multiplied by the normal time prior to adding to it, source 2 suggests an algorithmic procedure for evaluating fatigue allowances. Source 1 recommendations assume that a 4% basic fatigue allowance is added to all jobs and Table 4.1 allowances are additional fatigue allowances. Source 3 assumes that a 10% fatigue allowance, which also includes personal allowance, is given to everyone and Table 4.1 allowances are additional fatigue allowances. Source 2 requires adding allowances from all factors and then subtracting 25% from it.

Most of Table 4.1 allowances are based on what has worked well. The general practice has been that if an allowance does not create "problems," it is

Table 4.1: Summary of Widely Used Fatigue Allowances from the Three 1970s Sources

Category	Source 1 (%)	Source 2 (%)	Source 3 (%)
Physiological	0–17	2–20	0–40
Psychological	0–13	8–34	0–18
Environmental	0–30	5–19	0–53

considered acceptable. This "acceptable" not only means lower productivity and waste in some cases, but also inadequate in many cases. Since, at the moment, there is enough doubt about Table 4.1 values as these values cannot be defended on the basis of sound, rational, and scientific theories, we need to review and recommend new methods for determining the variable part of fatigue allowances. The following sections discuss methods that we consider objective and scientific. The methods are grouped under the categories physiological fatigue allowances, psychological fatigue allowances, and environmental fatigue allowances in keeping up with the widely accepted categories as shown in Table 4.1.

4.3.1 Physiological Fatigue Allowances

Metabolic energy expenditure rate, static load, strength, or muscular force, and heart rate are the most widely accepted physiological measures of work. These measures are known to provide separate and independent limits for human performance, and consequently fatigue allowances, in any task.

The main theme behind the methods based on static strength or muscular force is the fact that the maximum holding time (T_{max}) is dependent on the force exerted (f, expressed as the proportion of maximum force, F), and the fatigue developed in the relevant muscle group is a function of the ratio of time of holding to maximum holding time. This means,

$$\text{Fatigue} = \text{Function}\left[\frac{T_{task}}{T_{max}}\right]$$

where

T_{task} = The time of endurance of a muscle group in a task at a given level of exertion

T_{max} = The maximum endurance time for that level of exertion (it depends on the force exerted/max capability)

Using this concept, maximum holding time T (in minutes), as a function of holding force, f (in pounds), and maximum holding force, F, is given as

$$T\ (\text{minutes}) = \left\{\frac{1.197}{(f/F - 0.15)^{0.618}}\right\} - 1.2083$$

This relationship is based on two observations: (1) reduction in maximum strength occurs only if the holding force is more than 15% of maximum strength and (2) maximum holding force is normalized by the average of arm, leg, and torso lifting strengths (100 pounds). As the holding force, f, increases, muscle fatigue builds up, resulting in reduced muscle strength. Recovery from muscle fatigue has been expressed in terms of variable fatigue recovery allowance as

$$RA = 18\left(\frac{t}{T}\right)^{1.4} * \left(\frac{f}{F} - 0.15\right)^{0.5} * 100$$

Provided f/F is >15%, in the above equation, RA is the fatigue recovery allowance; t is the working period (holding time) in minutes; and T is the maximum holding time.

If a load of 50 pounds is lifted once every 10 min, fatigue rest allowance will be determined as follows:

$$T = \left\{\frac{1.197}{(50/100\ -\ 0.15)^{0.618}}\right\} - 1.2083 = 1.085\ \text{min}$$

Taking that it takes only 10 s to lift the load, RA, the variable fatigue allowance, would be

$$\text{RA} = 18\left(\frac{0.167}{1.085}\right)^{1.4} \times (0.5 - 0.15)^{0.5} \times 100 = 4.3\% \text{ (approximately)}$$

The resulting total allowance would be

$$= 5\% \text{ (personal allowance)} + 4\% \text{ (basic fatigue allowance)}$$
$$+ 4.3\% \text{ (variable fatigue allowance)} = 13.30\%$$

Methods that are based on *heart rate* take advantage of the fact that the heart rate is influenced by both static and dynamic work. One way to determine the upper limit of continuous work for an 8-h shift is to limit the workload such that within it the heart rate (or working pulse) returns to the resting level after approximately 15 min. *When the working pulse is 30 beats/min above the resting pulse, this limit is reached.* This means that working loads that elicit an increase of 30 beats/min in working pulse are the limits of continuous work performance.

It is, however, very unlikely that the working pulse will remain unchanged throughout the workday. The maximum heart rate varies throughout the day and the most frequent value lies between 130 and 140 beats/min, with an occasional much higher increase. In such cases, fatigue rest allowances should be determined such that the time-weighted average working pulse does not exceed 30 beats/min. Further, the frequency of extreme pulse increases (170/180 beats/min) should be less than 5%. Thus, for intermittent work, variable fatigue rest allowances for an 8-h shift (= 480 min) can be determined from the following relationship:

$$R_p + 30 = \frac{(T_r \times R_p + T_w \times W_p)}{480}$$

where $T_w + T_r = 480$ min; T_r is the rest time in minutes; T_w is the working time in minutes; R_p is the resting pulse/minute; and W_p is the working pulse/minute. Knowing the working pulse, fatigue rest allowance can be calculated.

Heart rate is generally higher when working intermittently as compared with working continuously at an even tempo. The above relationship is, therefore, also applicable to continuous work to determine variable fatigue rest allowances.

Metabolic energy expenditure (oxygen consumption)-based methods assume that it is possible to spend energy at a rate of 4 kcals/min (1 L of oxygen is equivalent to 5 kcal of metabolic energy). This is defined as the endurance limit (EL). Using the concepts of a standard metabolic rate and a basal metabolic rate, various relationships to determine variable fatigue rest allowances have been proposed in the literature. One such relationship is given as follows:

$$\text{Rest allowance (\% of total working time)} = \frac{(B - S)}{(B - 1.5)}$$

where B is the work energy expenditure in kcal/min; S is the standard working rate (5 kcal/min for men and 4.2 kcal/min for women), and 1.5 is the average energy expenditure rate.

The major shortcoming of models, such as above, is the assumption that 1/3 of the aerobic capacity (maximum volume of oxygen that can be consumed, equivalent to 15 kcal/min for men and 12.5 kcal/min for women) is the limit of continuous work performance for 8-h work performance. It has been demonstrated scientifically that it is not so. Further, it has been shown repeatedly that aerobic capacity is task dependent and varies widely. The limit of energy expenditure is not the same for all patterns of work, neither for all people nor for all groups of muscles. Environmental heat, muscular effort, and body weight also influence aerobic capacity. It would be, therefore, logical to look at the "total available energy" concept to determine the duration of variable fatigue allowances, particularly for dynamic work. A procedure that utilizes the total daily energy requirement concept is described here. The total energy requirement is adjusted for age and energy required for food ingestion. The calculation of basal (energy requirements while asleep) and leisure metabolism is based on the hours during which these activities are performed (default is 8 h). The work (shift) duration is also taken into consideration. It is assumed that there is a limit to the amount of metabolic energy available to produce work (one cannot consume unlimited amounts of food, for instance). Figure 4.1 shows the procedure for determining variable fatigue rest allowance.

To use the procedure, the following information is required:

1. Worker's gender (male or female)
2. Worker's age (years)
3. Worker's body weight (pounds)
4. Number of hours of sleep per day (hours)
5. Shift duration (hours)
6. Number of tasks performed during the shift
7. Time duration of each task (hours)
8. Metabolic energy requirement for each task (kcal/min)
9. An estimate of worker's physiological condition (high, average, or low)

   To increase the accuracy of the procedure, two additional pieces of information are needed:

10. Worker's aerobic capacity (mL/min/kg)
11. Average energy requirement/minute when not working or sleeping (kcal)

In the absence of these additional data, the procedure estimates variable fatigue allowances conservatively.

As shown in Figure 4.1, if the aerobic capacity of the worker is known then his/her physiological condition is calculated from Table 4.2. If the aerobic capacity of the worker is not known, a qualitative estimate (low, average, or high; smokers, for instance, are likely to have low aerobic capacity) is needed. Once the general physiological condition of the worker is known, the total amount of energy available to an 18-year old, per day, is determined from Table 4.3. Next, this total amount is adjusted for age. A decline of 1% per year, over 18 years, is used as the adjustment factor. A 10% reduction is made in this value to allow for energy required in digesting food.

Once these adjustments have been made, basal metabolism and energy expenditure during the leisure period are calculated using a value of 0.9 kcal/kg of body weight/h of sleep (corrected for metabolic savings in sleep). To determine the energy requirement during leisure time, either the actual value (in kcal/min) is used or a default value of 1.5 kcal/min is used (this value represents energy expenditure during a typical leisure time sedentary task).

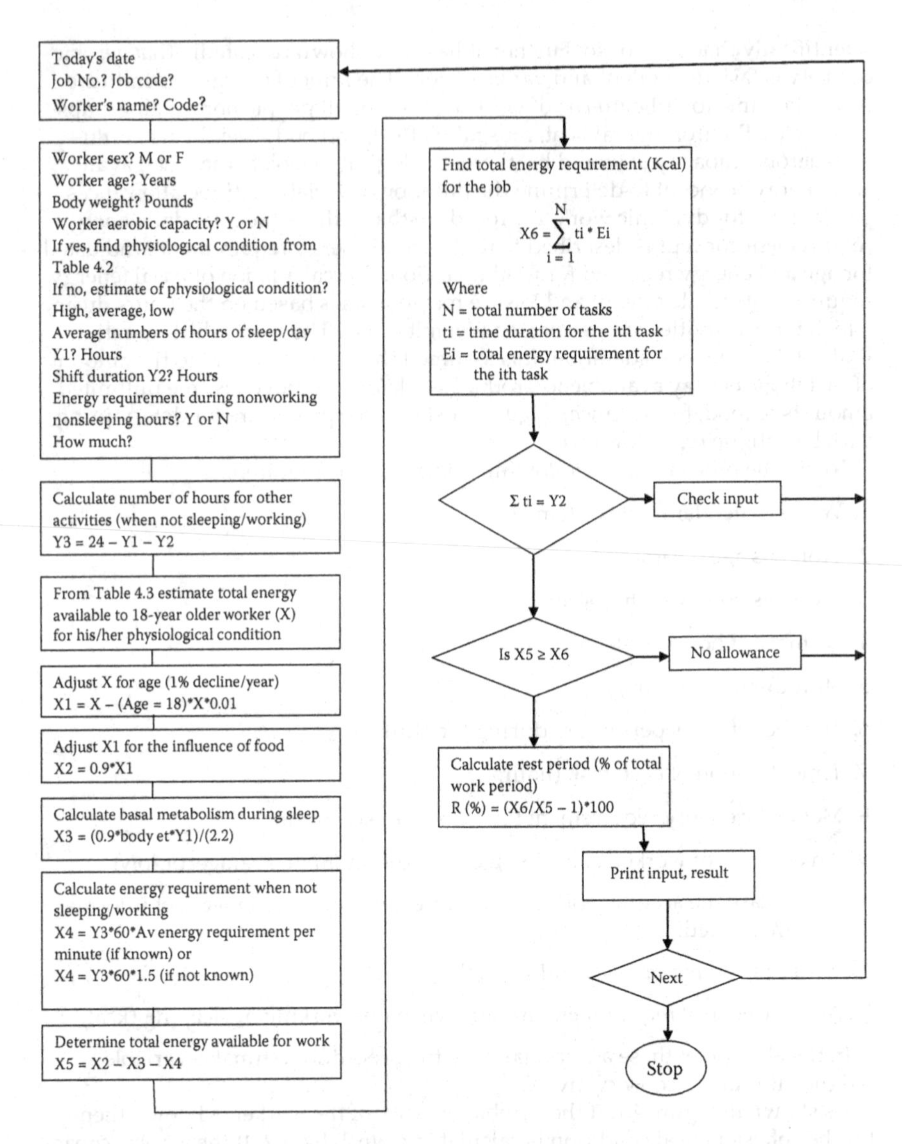

Figure 4.1 Rest allowance determination procedure based on metabolic energy expenditure.

The basal and leisure metabolism are subtracted from the adjusted total energy available to determine the amount available for producing work. Next, the total energy requirement for the job is determined as follows:

$$\text{Total energy requirement for all jobs (kcal)} = \sum_{i=1}^{N} t_i E_i$$

where t_i is the time (in hours) for the ith job; E is the energy requirement in kcal; and N is the total number of jobs carried out during the shift. Next, the rest period, as a percentage of the shift duration, is determined as follows:

Table 4.2: Range of Expected Maximal Aerobic Capacity by Gender and Age

	Maximal Oxygen Uptake (mL/kg/min)				
Age (Years)	Low	Fair	Average	Good	High
Men					
20–29	25	25–33	34–42	43–52	52+
30–39	23	23–30	31–38	39–49	49+
40–49	20	20–26	27–35	36–44	44+
50–59	18	18–24	25–33	34–42	42+
60–69	16	16–22	23–30	31–40	40+
Women					
20–29	24	24–30	31–37	38–48	48+
30–39	20	20–27	28–33	34–44	44+
40–49	17	17–23	24–30	31–41	41+
50–59	15	15–20	21–27	28–37	37+
60–69	13	13–17	18–23	24–34	34+

Table 4.3: Total Energy Available/Day (kcal) for an 18-Year Old

Gender	Aerobic Capacity	Low	Fair	Average	Good	High
Male	Known	2100	2575	3050	3225	4000
	Unknown	2100	–	3050	–	4000
Female	Known	2100	2325	2550	2775	3000
	Unknown	2100	–	2550	–	3000

$$\text{Rest}(\%) = \left[\left(\frac{\text{Total energy requirement for the job}}{\text{Total adjusted energy available for work}}\right) - 1\right] \times 100$$

The procedure described here is applicable to any shift duration and any combination of dynamic physical tasks.

To simplify the usage of the procedure, an interactive program (written in *BASIC*) has been developed and is listed in Appendix.

Now we show the application of the program for an actual case. The differences in rest allowance determination when the aerobic capacity is known and when it is not known are also shown.

The individual involved is a man with the following particulars:

Age	29 years
Body weight	145 lbs
Aerobic capacity	55 mL/kg/min
Hours of sleep/day	8
Shift duration	8 (case 1) and 12 (case 2)
Energy requirements during leisure time (kcal/min)	1.7
Energy required for the job (kcal/h)	237.6 (case 1) and 206 (case 2)

Note: 237.6 = 60 min × 3.96 kcal/min for case 1; 206 = 60 min × 3.44 kcal/min for case 2.

Two different cases are examined: case 1, 8-h shift; case 2, 12-h shift. Table 4.4 shows the program output for the two cases. In neither case is a rest allowance needed. The total energy requirement for the job is quite close to the total energy available for work.

Table 4.5 shows the program output for a different situation for an 8-h shift when the aerobic capacity is known and when it is unknown and has to be estimated.

Encountering highly repetitive tasks using the upper extremities, such as fingers and wrists, is also quite pervasive in industry. There is no reliable information available to date to determine fatigue allowances in such cases. Anecdotal evidence suggests that in some operations, such as typesetting or keyboard operations, 11,000–12,000 finger operations per hour are not problematic. For hand and wrist operations, problems typically precipitate when the number of strenuous motions (motions requiring excessive force or limb deviation) exceed 10,000 per shift. These numbers may be used as a crude marker for determining fatigue allowances. It helps if the motions are spread over the day, evenly.

4.3.2 Psychological Fatigue Allowances

Information overload, concentration, monotony in vigilance tasks, and repetition are some of the factors that lead to fatigue in cognitive tasks. Suitable fatigue allowances are therefore needed for such tasks. As shown in Table 4.1, these allowances vary considerably and also, as in the case of physiological fatigue allowances, are based on empirical methods. A number of procedures are cited in the published literature for prescribing variable psychological fatigue allowances. They range from: (a) those that relate the fatigue allowance with performance decrement with time, (b) those that measure pause time or recovery time, (c) those that involve objective or subjective measurement of fatigue, and (d) those that use empirical methods.

In recent years, a number of scientific relationships between performance and factors affecting performance have been proposed. While these relationships do not always directly provide psychological fatigue rest allowances, rest periods can be determined by assessing performance loss resulting from variation in factors that affect performance.

For *target detection tasks*, the following relationship may be used:

$$\%\ \text{Targets detected} = 81\, C^{0.2} L^{0.045} T^{-0.003} A^{0.199}$$

where L is the background luminance (in foot-lamberts), C is the contrast, T is the observation time (in seconds), and A is the target visual angle (in minutes of arc). C is multiplied by 1.5 for spatial or temporal uncertainty, by 2.78 for moving targets, and by 2.51 for outside conditions. The percentage of targets detected based on the individual's ability may be used to define variable fatigue allowances. On this basis, rest allowances, as a percentage of total time for the task, are given as follows:

- 0% for 95% or higher rate of detection (task with no significant problems)
- 2% for 50%–95% detection (fine or exacting work)
- 5% for below 50% detection (very fine or very exacting work)

Mental strain or mental fatigue is very difficult to measure directly and objectively. Measures of performance, such as output, are easily influenced by motivation and job monotony. Despite these problems, a number of relationships have been developed. Yet, which of these relationships should be used is still

Table 4.4: Computer Program Output for the Sample Case

Date:	7/17/2015
Job number:	Example
Job code:	Case 1
Worker name:	Mr. X
Sex:	M
Age (years):	29
Weight in pounds:	145
Aerobic capacity (mL/kg/min):	55
Physiological condition computed from the table:	5
Hours of sleep per day:	8
Shift duration (hours):	8
Energy available for 18-year old from Table 4.3 (kcal):	4000
Energy required when not sleeping and not working (kcal):	816
Total energy available for work (kcal):	1913.45
Total number of jobs:	2
Total duration of job (hours):	Energy required for job (kcal/h)
7	237.6
1	102
Total energy requirement of the job (kcal):	1765.2
Rest period as a percent of shift duration:	0
Program output	
Date:	7/17/2015
Job number:	Example
Job code:	Case 1
Worker name:	Mr. X
Sex:	M
Age (years):	29
Weight (pounds):	145
Aerobic capacity (m/kg/min):	55
Physiological condition computed from the table:	5
Hours of sleep per day:	8
Shift duration (hours):	12
Energy available for 18-year old from Table 4.3 (kcal):	4000
Energy required when not sleeping and not working (kcal):	408
Total energy available for work (kcal):	2321.45
Total number of jobs:	2
Total duration of job (hours):	Energy required for job (kcal/h)
10	206
2	102
Total energy requirement of the job (kcal):	2290
Rest period as a percent of shift duration:	0
Program output	

Table 4.5: Program Output with Known and Unknown Aerobic Capacity

Aerobic Capacity Known	
Date:	7/17/2015
Job number:	Example
Job code:	Case 1
Worker name:	Mr. X
Sex:	M
Age (years):	35
Weight (pounds):	150
Aerobic capacity (mL/kg/min):	44
Physiological condition computed from the table:	4
Hours of sleep per day:	7
Shift duration (hours):	8
Energy available for 18-year old from Table 4.3 (kcal):	3525
Energy required when not sleeping and not working (kcal):	810
Total energy available for work (kcal):	1393.63
Total number of jobs:	3
Total duration of job (hours):	Energy required for job (kcal/h)
2	180
3	200
3	200
Total energy requirement of the job (kcal):	1560
Rest period as a percent of shift duration:	11.9397
Program output	
Aerobic Capacity Unknown	
Date:	7/17/2015
Job number:	Example
Job code:	Case 1
Worker name:	Mr. X
Sex:	M
Age (years):	35
Weight (pounds):	150
Aerobic capacity (mL/kg/min):	44
Estimated physiological condition:	3
Hours of sleep per day:	7
Shift duration (hours):	8
Energy available for 18-year old from Table 4.3 (kcal):	3050
Energy required when not sleeping and not working (kcal):	810
Total energy available for work (kcal):	1038.8
Total number of jobs:	3

(Continued)

Table 4.5: (*Continued*) Program Output with Known and Unknown Aerobic Capacity

Aerobic Capacity Unknown

Total duration of job (hours):	Energy required for job (kcal/h)
2	180
3	200
3	200
Total energy requirement of the job (kcal):	1560
Rest period as a percent of shift duration:	50.1726
Program output	

open to speculation. *The criticality of the task should also be considered; not all tasks are equally important even though they may be equally complex.* What can be stated is that as the task complexity increases, the need for higher allowances arises. Moreover, rest pauses should increase in frequency. Anecdotal information suggests that errors increase at an unacceptable rate after a concentrated effort of 20 min. *This means that a rest pause may be prescribed after a working duration of 20 min.* In general, one can expect a decline in performance ranging from 2% to 4% per hour of work, depending on the task complexity. Rest pause duration may be determined from the following relationship:

$$\text{Variable fatigue allowance (\%)} = 100\left\{1-(0.0068T^{0.114}L^{2.065}+1)^{-1}\right\}$$

where T is the duration of work time and L is the task difficulty. L may be measured by indicators such as the number of problems solved incorrectly per minute (higher this number, lower would be the allowance). Other measures of task difficulty, such as errors, may be used with caution as the above relationship was developed using only mental arithmetic.

Decline in performance should also be expected for *monotonous* tasks. In general, a decline of 2% per hour should be expected; the decline will be higher if the task requires intense mental and visual concentration. *In the absence of any concrete scientific recommendations available in the published literature, an allowance similar to the one for mental strain may be used.*

4.3.3 Environmental Fatigue Allowances

Climatic conditions, noise, vibration, and illumination are the major environmental factors that are known to cause performance-influencing stress. In this section, we focus on heat, noise, illumination, and vibration.

Work in modern industry is quite different from work in industries in the 1900s. There are fewer situations where workers are exposed to adverse situations. Obviously, outdoor work is susceptible to atmospheric conditions. Even indoors, we occasionally encounter situations that add *heat stress* problems; for instance, while working in steel mills and foundries. The worker in such situations not only has to sustain the heat load resulting from climatic conditions but also the metabolic heat gain resulting from performing physical work (metabolic work load is expressed in kcal/min and is measured by measuring oxygen consumption in liters/minute; see Chapter 9 for details). Thus, both external and internal heat loads need to be integrated in order to determine variable fatigue allowances. Continued exposure to internal and external heat loading leads to elevated body core temperature. The current scientific thinking is that the elevation in the core temperature should be limited to 1°C or 1.8°F. Using this limit as the limit of working time, rest allowances can be calculated as follows:

$$\text{Variable fatigue allowance (\%)} = e^{(-41.5+0.0161W+0.497WBGT)}$$

where W is the metabolic energy expenditure rate (kcal/min) and WBGT is the wet bulb globe temperature. The WBGT is an integrated index accounting for the effect of radiative and convective heat loads and humidity and can be readily measured using commercially available equipment. The above relationship, thus, integrates both internal and external heat loads.

Noise exposure may not necessarily lead to deterioration in performance, unless it affects job-related communication. It does affect worker health, however. With that in mind, the Occupational Safety and Health Administration (OSHA) in the United States has limited worker exposure to noise to 90 decibels on A scale (90 dBA) for an 8-h work shift. The noise exposure is allowed to increase by 5 dBA for every halving of the exposure time. So, for instance, a worker can be exposed to continuous noise of 110 dBA for a period of 30 min. The permissible exposure time to different noise levels can be calculated from

$$\text{Exposure time T (hours)} = \left[32/2^{((L-80)/5)}\right]$$

where L is the noise level in decibels on A scale (dBA).

The rest allowance is determined from the following relationship:

$$\text{Variable fatigue allowance} = 100\ (\text{Dose} - 1)$$

where the dose is determined by the expression

Dose $= T_{e1}/T_{p1} + T_{e2}/T_{p2} + , \ldots, +T_{en}/T_{pn}$, where T_e is the actual exposure time at a given noise level and T_p is the OSHA permissible time exposure limit at that noise level. For the entire workday exposure to be OSHA compliant, the dose must not exceed 1 (≤ 1); if it is greater than 1, rest allowance would be required as above.

There is significant scientific evidence to suggest that task performance improves with task *illumination*. As tasks vary in complexity, illumination levels become more crucial for highly complex tasks that require a great deal of visual concentration (e.g., looking for surface cracks on a machined part). Two different models are provided to calculate task performance times, one for complex tasks requiring a great deal of visual concentration and the other for regular tasks:

$$\text{Performance time (seconds)} = 25.9 - 7\log(FC) + 1.45\log(FC)^2$$

for complex tasks such as precision inspection

$$\text{Performance time (seconds)} = 251.8 - 33.96\log(FC) + 615\{\log(FC)\}^2 - 0.37\{\log(FC)\}^3$$

for regular tasks such as reading

where FC is the illumination level in foot-candles. As the illumination level decreases, the performance time increases necessitating a higher allowance. The Illuminating Engineering Society (IES) has made illumination level recommendations for a variety of tasks. These recommended illumination levels could be plugged in the models above and a desired performance time obtained. A decline in performance time resulting from reduced illumination can then be tied to rest allowances:

- Decline in illumination level of 3% from the IES recommended level or a decline in performance of up to 3% from the desired performance time would result in no fatigue allowance.
- Decline between 3% and 5% in illumination level or performance would result in a 2% fatigue allowance.
- Decline between 5% and 10% in illumination level or performance would result in a 5% fatigue allowance.
- Decline between 10% and 15% in illumination level or performance would result in a 10% fatigue allowance.

These recommendations have been synthesized from the published literature.

For *vibration* exposure, either segmental or whole body, there are no scientific relationships that allow determination of variable fatigue allowances. There are, however, recommendations made by different organizations to limit exposure to vibration. For instance, the American Conference of Governmental Industrial Hygienists (ACGIH) recommends a vibration exposure of less than 4 m/s^2 for exposure of 4–8 h. For every 2 m/s^2 increase in vibration acceleration, the exposure limits are reduced by half (2–4 h for 6 m/s^2; 1–2 h for 8 m/s^2; <1 h for 12 m/s^2). *The ACGIH also recommends no vibration exposure for every 1 h of continuous vibration exposure.*

5 Work Sampling

5.1 THE CONCEPT OF WORK SAMPLING

At the core of every business activity is the pursuit of increased profits. Profits are the difference between revenue and cost. Assuming that the total revenue remains constant, profit then becomes a function of cost. An organization that possesses flat revenues (generally a symptom of very little pricing power) can still maximize its profitability by controlling costs.

The cost parameter is almost always a function of productivity. Thus, there is a correlation that profitability is driven by productivity. A manager who desires to raise profits will always seek to maximize productivity. The productivity of any task can be determined by computing the percentage of time that is spent in value adding activities as opposed to that spent in nonvalue adding activities. A process often referred to as work sampling can measure this. Work sampling was first applied in the British textile industry. Later, the technique made its way to the United States.

Work sampling may be defined as a statistical technique used to determine the amount of time spent by an employee in performing a variety of different activities. It is used to investigate the proportion of total time devoted to various activities that constitute a job or work situation. The results of work sampling are used to determine machine and personnel utilization, allowances applicable to the job and production standards. Work sampling frequently provides the same information faster and costs considerably less than time study. Work is analyzed by taking a large number of observations at random intervals of time. The accuracy of the data determined by work sampling depends on the number of observations and the period over which the random observations are taken. Work sampling studies rely on a large number of observations obtained at random intervals of time. The ratio of observations of a given activity to the total number of observations is an approximation of the total percentage of time that the process is in that state of activity.

Work sampling is based on the laws of probability and probabilistic distributions. Over the course of the study, several parameters, such as the state or condition of the object of study, are carefully observed and classified into different classes of activity that represent that particular work situation. The total work activity being studied is then analyzed.

The concept of work sampling is based on the laws of probability. These laws state that a smaller number of random occurrences will tend to follow the same distribution as the entire populations. Therefore the sample size (n) will approximate the population (size = N), where ($n \ll N$). This implies that inferences drawn from observing a process being performed by an operator can be confidently applied to the entire process. The accuracy increases with the number of observations. They are directly proportional. Thus, as "n" approaches "N," the accuracy rises.

5.2 ADVANTAGES OF WORK SAMPLING

Work sampling provides a variety of advantages over stopwatch time study or any other work measurement method. These advantages are enumerated as follows:

1. Work sampling results are generally unbiased due to the fact that the employee is not being observed directly
2. Given the fact that work sampling is based on statistical analysis and the law of probability, it can be disturbed at any time without any adverse impact on the outcome of the study

3. The need to have experts conduct a work measurement study can be done away with as far as work sampling is concerned. Anyone with limited amount of training is capable of performing such a study
4. In comparison with time study, work sampling is considered to be far more economical and efficient in terms of amount of time consumed. This can be attributed to the fact that several employees can be studied simultaneously
5. Work sampling can provide vital information about employees and machines in less time and lower cost

5.3 DISADVANTAGES OF WORK SAMPLING

Despite having several distinct advantages, work sampling has a few drawbacks in terms of its field of application. The said drawbacks are enumerated as follows:

1. In terms of economy, work sampling is not economical for short cycle jobs, since the number of observations from such a study is not large enough to try and approximate a population
2. Work sampling does not subdivide the activities and allow for delays or smaller classifications of work
3. An employee under observation could deviate from one's normal work routine or alter one's pace of work. If such is the case, then it may render the final result of the work sampling study quite inaccurate and undependable
4. If the number of observations in a work sampling study is not large enough, they may interfere with the underlying statistical framework, thus rendering the entire study ineffective
5. The rate at which an employee performs one's job (the pace of work) is not accounted for by work sampling. This could be a drawback that would hinder the accuracy of the study

5.4 PROCEDURE FOR CONDUCTING A WORK SAMPLING STUDY

A work sampling study can be designed by adhering to the following steps:

1. *Problem definition*: The problem that is to be studied needs to be clearly defined. This includes stating the objectives and describing each element of the study in detail. Determine the information that is required. This step can be simplified by breaking down the operation to be studied into its constituents and determine what needs to be studied when and in what amount of detail.
2. *Sell the study*: This step involves getting the permission of management as well as that of the object of observation (namely the employee being observed). Both parties need to be convinced of the need and the utility value (in terms of potential benefits to be obtained) of the study. Using statistical analogies and quantitative data will serve to corroborate the "sales pitch."
3. *Make an observation recording form*: This step involves creation of an appropriate form to be used for recording the observations to be made during the course of the study. Given the varying nature of work, such forms must be custom made individually for each work sampling study. The design of the form is a function of the number of workstations and operators to be studied and the types of activities under observation.

Table 5.1: A Typical Work Sampling Form

Date	Work sampling data collection form										Page of
Period of study							Activity category (AC) 1. Keypunch 4. Telephone 2. Writing 5. Walking 3. Filing 6. Conversation				
Observer											
Department											
AC = activity category; PR = performance rating											
Subjects											
Observation date and time	Smith		Wang		Jones		Zhang		Willis		Notes
	AC	PR	AC	PR	AC	PR	AC	PR	AC	PR	

4. *Select the frequency of observation*: This step involves quantifying several parameters of the study. Information needs to be finalized in terms of the number of observations to be measured, the number of observers involved in the study, the number of days and shifts required to complete the study, ascertaining the time and the route to be followed by the observer, and the total time limit (maximum amount of time allowable) of the study. It is important to remember to take as many observations as possible in order to enhance the accuracy of the study. The number of observations can be predetermined based on the level of accuracy desired by using the formula described in the following pages.
5. *Evaluate methods to reduce bias*: Sometimes, inefficient components of a method may not necessarily be self-evident in a work sampling study. For instance, work being performed during equipment downtime can also be performed while the machine is working. Similarly, a work sampling study will not specify if the operators are pacing themselves or operating at optimal speed. This issue has already been touched upon in the disadvantages section of work sampling.

It should be noted that no work sampling studies should be undertaken without the consent of the person being studied. It is important to meet with the object of the study prior to the study and discuss each element to be observed and recorded in depth. This is especially important in a situation where two or more observers study the same person, from the point of view of consistency.

An example of a work sampling form is depicted in Table 5.1. Another example is depicted in Table 5.2. The difference in the two forms is obvious. In Table 5.2 for instance, the form separates productive work from nonproductive work and the corresponding percentages are noted. Moreover, Table 5.2 does not distinguish between operators. Rather, it is an aggregate consideration of all the work done in a manufacturing facility.

Table 5.2: Work Sampling Study Form Classifying Aggregate Activities

Work sampling study

XYZ Manufacturing By: ________________

Number working this study: ____________ Date:____________ Remarks:______________________________

No.	Random time	Productive activities					Nonproductive activities						Total Observations	% Productive	% Nonproductive
		Machining	Weld	Elect	Carpent	General labor	Get Tools	Grind tools	Wait job	Wait Crane	Personal	Idle			

5.5 ACCURACY AND PRECISION OF WORK SAMPLING

There is a considerable amount of variability that is inherent in work measurement. It is imperative to define the terms accurately and precisely in order to provide effective results through work sampling.

Accuracy may be defined as the degree of bias inherent in any measuring activity. Bias is a measure by which a long-range mean of observations deviate from the true value of the quantity being measured.

Precision, on the other hand, is the extent to which the measured value of a given quantity can be reproduced for any given period of time without any regard for the true value of that quantity. A work measurement study if flawed can be precise but inaccurate. This implies that it could stray from the true value of a consistent quantity and always reproduce that inaccuracy with a high degree of consistency.

The only way to prevent bias is by structuring the work sampling study carefully and properly. The execution of work sampling needs to be undertaken with just as much diligence.

Bias could be introduced into a work sampling study due to a variety of different reasons like those stated below:

1. Precisely defining the population that needs to be sampled
2. Various states of activity that constitute the work are ambiguously defined
3. The person performing the work sampling study exhibits a considerable amount of latitude in terms of choosing the moment of observation
4. The method of selecting observation times is inconsistent and perhaps defective
5. The worker, or in other words the person that is the object of the study, could anticipate the moment at which the study is undertaken and is capable of altering the nature of the activity that they perform. Such an alteration generally would constitute a significant deviation from the true nature of the operations

It has to be kept in mind that during the design stage of a work sampling study, a period should be selected that will tend to avoid an unusual situation. This will enable the work sampling to accurately represent normal working conditions. Besides, the period of study should be at least as long as the longest period of any cyclical tendency that a constituent work activity may exhibit. It goes without saying that studying the population forms an estimate. This population should be similar and characteristic of the period to which the estimate is applied in the future.

So, what is the difference between conducting an actual time study and work sampling? An important distinction needs to be made at this juncture. Chapter 3 deals with the calculation of sample size as well. Detailed formulas have been presented in Chapter 3, which deal with the calculation of sample size that is essential in performing time study. The Chapter 3 formulas deal with computing sample size when some of the time values are already known (this is taken to be the value of X). It will be appreciated that the value of X is measured not counted; the unit of measurement in that case being time that is measured in seconds. Thus, X can be likened to a variable.

On the other hand, in work sampling, the task is to determine what percentage of time an operator is busy performing a certain task as against being idle. The actual time values are not known, so the occurrences are counted and not timed. This is unlike an actual time study. In this context, work sampling data can be likened to an attribute. For readers that are not familiar with the terminology, definitions of variables and attributes are presented here:

What is a variable: A variable is something that can be measured. For instance, length, volume, and weight are variables because they are measurable. Each of the aforementioned examples has units of measurement.

What is an attribute: Attributes are counted, not measured. For example, determining what percentage of items meet certain quality standards is accomplished by counting the number of items, not measuring them.

It has been mentioned at the beginning of this chapter that the concept of work sampling is based on the laws of probability. Probability accounts for the likelihood of the occurrence of a certain event. This is generally quantified using a distribution referred to as a binomial distribution. The objective of work sampling is to compute the value of $\bar{p}$. "p" is the percentage occurrence of an element being observed expressed as a decimal. The parameters p and n (n = number of observations in a sample) are used in binomial distribution. s is the notation for standard deviation and is used in the normal distribution. The value of s can be computed using the following formula:

$$\sigma = \sqrt[1]{\frac{\bar{p}(1-\bar{p})}{n}}$$

68% of the time an observation would not deviate from the mean in a normal distribution by more than ±s. It also can inform about the probability "a" associated with more than 1s.

The percentage of area under the normal curve "a" or the selected level of confidence "a" between the arithmetic mean of the distribution and a certain point to the left or to the right of the mean is tabulated in Table 5.3.

Mathematically, the number of observations required to be obtained in order to achieve a specific degree of accuracy can be computed using the following formula:

$$n = \sqrt{\frac{C^2\bar{p}(1-\bar{p})}{\sigma^2}}$$

In the above formula, "C" is the confidence level read from Table 5.3.

The rule of thumb, however, is to record as many observations as possible in order to assure a high degree of accuracy.

An alternative to using the aforementioned formula is to use the alignment charts to determine the number of observations required in order to ensure a certain degree of precision. An example of an alignment chart is presented in Figure 5.1.

Table 5.3: Levels of Confidence Used to Determine Sample Size

C (+ and –)	A
1.000	0.68
1.645	0.90
1.960	0.95
2.576	0.99

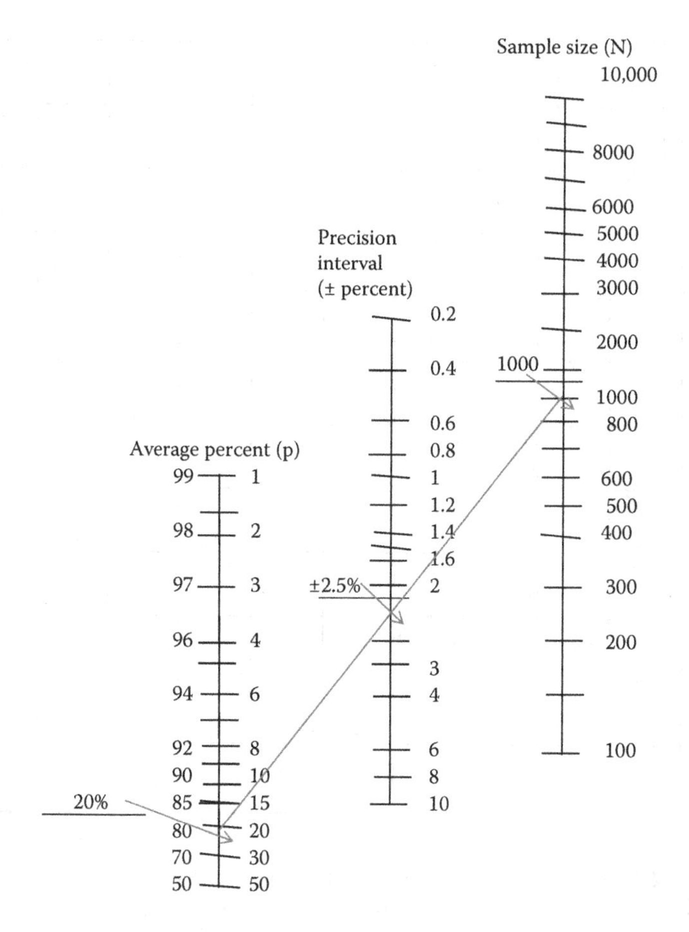

Figure 5.1 Alignment chart to determine the sample size for work sampling.

Another method to calculate the sample size, thus an alternative to alignment chart, is to use the diagram presented in Figure 5.2. The chart uses a predetermined value of standard deviation in order to compute the sample size. The figure suggests that if the sample size 1was to remain unchanged, the standard error would be inversely proportional to $\bar{p}$.

5.6 CONTROL CHARTS IN WORK SAMPLING

Control charts are used in work sampling studies in order to enable the analyst to plot the results of the study on an ongoing basis. If a data point falls outside the control limits, it likely indicates the presence of some unusual or abnormal condition during that part of the study.

3 sigma (3s) is generally used to determine the upper and lower control limits of the control chart. This implies that there are only three chances in 1000 that a point will fall outside the limits due to a chance cause. The assumption can be safely made that if and when a point does fall outside the limits on either side, there is an assignable reason for it.

p charts (percent defective) are usually used in this context. p charts are used to plot the percent defectives in any sample in order to ascertain if the

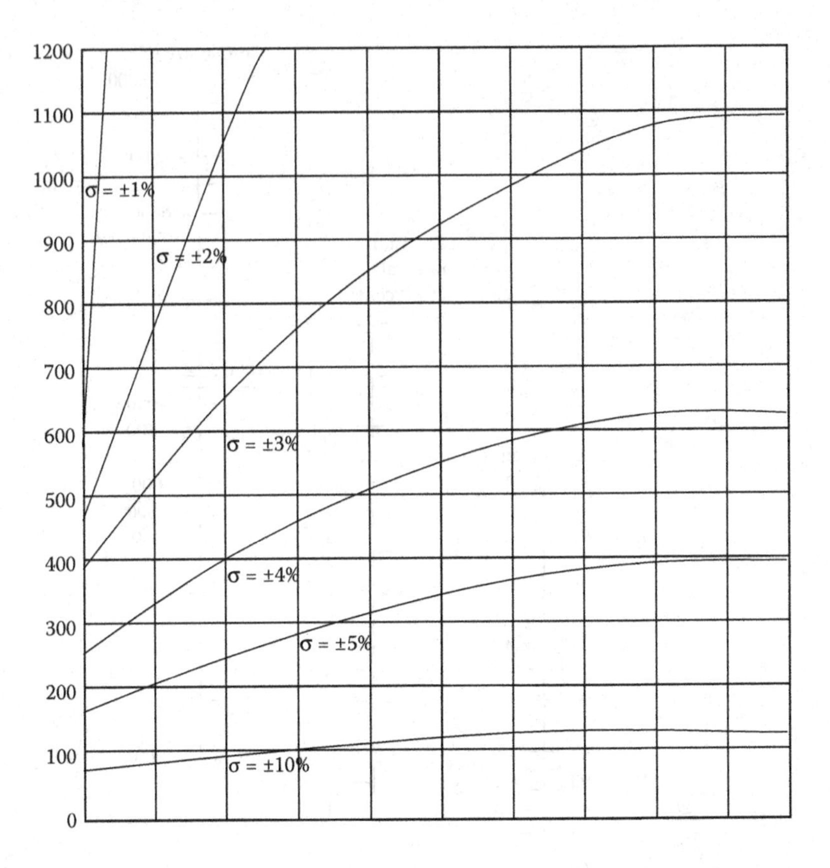

Figure 5.2 An alternative method to determine sample size in work sampling: Number of observations required to obtain a specific percent occurrence within absolute limits of error.

underlying process is in control. To plot a p chart, it is first essential to determine the value of s and $\overline{p}$. $\overline{p}$ is the average value of observations and is set as the mean level for the control chart. The value of s can be computed using the formula

$$s = \sqrt{\frac{\overline{p}(1-\overline{p})}{n}}$$

The bounds for the control chart (upper and lower control limits) can be determined using the following formula:

$$\overline{p} \pm 3\sqrt{\frac{\overline{p}(1-\overline{p})}{n}}$$

Another method of expressing the above equation is to write the following:

$$\overline{p} \pm 3s$$

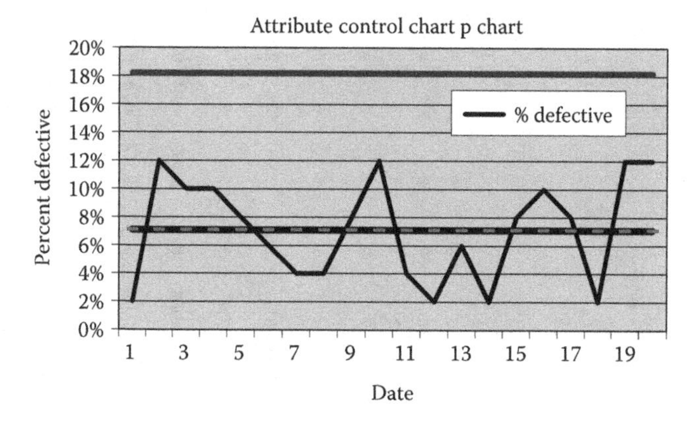

Figure 5.3 Example of an attribute control chart (p chart).

The above equation would yield two distinct values. The upper control limit would constitute the higher level and the lower control limit would be the lower value obtained from the equation.

An example of a "p" chart is depicted in Figure 5.3. The chart is derived from raw data presented in Table 5.4. Assume that 20 samples were inspected at random and the objective is to look for defectives. The sample size is 50. Application of the aforementioned formulas yields values for the mean percent defective, and for upper as well as lower control limit.

Therefore, in this particular case, an item is said to be defective if it is either miscoded or has the wrong price assigned to it. For sample 1, 1 out of 50 items is miscoded and none is wrongly priced. This counts as a total of one defective. The percent defective is computed by dividing the total number of defectives by sample size, which in this case equates to 2%.

Continuing this process across all samples, we obtain the following:

- Individual values of percent defective for each sample
- Finding the average if individual percent defective values yields the aggregate value of average percent defective that in this case corresponds to 7%
- The value of standard deviation is computed by applying the formula presented above. In this case, the value of standard deviation corresponds to 3.7%
- The value of control limits is computed similarly by using the corresponding formula
- The corresponding value of upper control limit in this case corresponds to 0.182% or 18.2%
- The corresponding value of lower control limit in this case is a negative number. Since there is no such thing as a negative percent defective, this value is rounded up to 0, which overlaps with the x-axis

On operations that are repetitive in nature and have short cycle times, elemental data or motion time data are preferred over work sampling. Given its reliance on a large number of observations, work sampling is preferred for operations with long cycle times. As mentioned before work sampling can be used to determine the productivity of an operator as well as their nonproductivity expressed

Table 5.4: Raw Data and Limit Values for an Attribute Control Chart (p Chart)

Attribute Control Chart																					Total
July	1	2	3	4	5	6	7	8	9	10	11	12	13	14	15	16	17	18	19	20	
	50	50	50	50	50	50	50	50	50	50	50	50	50	50	50	50	50	50	50	50	1000
Miscoded	1	3	3	3	1	1	1	2	1	4	0	0	2	0	2	3	2	1	4	3	
Wrong price	0	3	2	2	3	2	1	0	3	2	2	1	1	1	2	2	2	0	2	3	
Total defective	1	6	5	5	4	3	2	2	4	6	2	1	3	1	4	5	4	1	6	6	71
Percent defective	2%	12%	10%	10%	8%	6%	4%	4%	8%	12%	4%	2%	6%	2%	8%	10%	8%	2%	12%	12%	0.071
Avg defective	7%	7%	7%	7%	7%	7%	7%	7%	7%	7%	7%	7%	7%	7%	7%	7%	7%	7%	7%	7%	
Std deviation	0.037	0.037	0.04	0.037	0.037	0.037	0.04	0.04	0.04	0.04	0.037	0.037	0.037	0.04	0.04	0.037	0.037	0.037	0.037	0.04	0.04
UCL	0.182	0.182	0.18	0.182	0.182	0.182	0.18	0.18	0.18	0.18	0.182	0.182	0.182	0.18	0.18	0.182	0.182	0.182	0.182	0.18	0.18

as idle time. The productivity of an operator can be measured by computing the value of the so-called "performance index." This value can be computed by using the following formula:

$$\frac{(\text{Number of pieces produced during the day})(\text{standard time per piece in minutes}) \times 100}{\text{Number of hours during the day}}$$

5.7 ALLOWANCES

Despite numerous advantages offered by the technique of work sampling in defining and ascertaining productivity of an operation, the field of establishment of allowances remains sorely ignored. Allowances are regrettably estimated or "guessed" during the application of time study.

To determine allowances using work sampling, it is crucial to standardize methodologies first. The standardization of methods is especially pertinent in the case of work sampling studies. This is due to the fact that work sampling has almost always been used together with work simplification in order to diagnose the problem and determine its cause. Work sampling will tend to expose the fault, but will not suggest any correction. It is essential, however, in the case of setting allowances to first standardize the method itself.

When designing a work method it is suggested that allowances be made to account for "personal time" during which personnel may rest and recoup their energy. This should be included as a policy matter. Moreover, items such as "inspection time" should be increased in order to ensure higher quality of finished products.

Productivity can be expressed in terms of machine utilization. This is defined as the total amount of time that a certain piece of machinery is actually being utilized. It is clear that the ultimate goal of any manufacturing manager is to maximize machine and equipment utilization. Table 5.5 depicts a work sampling summary sheet. It will be appreciated that the percentage of time consumed by various delays point to facets of work that could be significantly improved through methods study. This would lead to optimization of machining time and thus machine utilization.

Table 5.6 depicts a summary of work sampling study to compute unavoidable delay allowances on a variety of operations, such as bench, bench machine, and spray operations. It will be observed that the unavoidable delay allowance was about 0.949% in this case.

5.8 PERFORMANCE AND COST EVALUATION PROGRAM

PACE is a mnemonic word used to describe the "performance and Cost Evaluation Program" used by Norair corporation in the 1950s. Management instituted this program in order to help supervisors ascertain the productivity of their employees working together as a group.

PACE is a work sampling technique that considers the following factors to determine employee productivity:

1. The number of employees assigned to a task
2. The amount of idle time
3. The amount of time employees are outside the assigned work area
4. An estimation of the performance rating that should be applied to the entire group

Table 5.5: Work Sampling Summary Sheet Depicting Each Machining Activity as a Percent of Total Machining Time

Work sampling summary

Date: ____________

Observer: ____________ Remarks: ____________

Machine	Cutting	Setup	Machine idle	Wait for crane	Inspection	Aid inspection	Wait for tools	Wait for tool problem	Conference with other shift	Tool handling	Obtain or grind tools	Confer with fman/inspc	Wait for job	Clean table and chips	Miscellan	Operator absent	Total time
20′ VBM	101	7	14	2	3	0	1	0	2	37	5	3	0	0	6	35	216
10′ VBM	102	34	14	15	3	1	1	0	1	28	5	1	7	4	0	0	216
28′ VBM	119	34	10	5	5	2	0	0	0	18	2	1	2	0	0	18	216
12′ VBM	109	24	12	13	6	1	0	0	3	26	6	2	3	3	2	6	216
16′ Planer	127	17	6	9	2	0	0	0	0	22	0	2	15	0	4	12	216
8′ IMM	64	18	17	16	3	0	0	0	2	30	7	3	0	0	28	28	216
16′ VBM	147	19	10	14	3	1	0	0	0	15	2	0	0	1	1	3	216
14′ Planer	140	8	5	7	2	0	0	0	2	17	3	0	3	0	11	18	216
72″ Lathe	99	13	12	7	3	0	0	0	1	32	8	2	0	0	3	36	216
96″ Lathe	89	9	29	18	11	1	0	0	2	29	8	3	4	0	3	10	216
96″ Lathe	109	14	12	8	10	0	3	3	0	32	9	8	2	0	1	5	216
160″ Lathe	72	34	13	14	6	2	1	0	4	21	3	3	1	1	4	37	216
11.5′ Planer	106	35	11	10	4	0	0	0	1	11	4	5	3	2	8	16	216
32′ VBM	151	23	8	7	1	0	0	0	1	10	2	1	5	3	5	0	216
Total	1535	289	173	145	62	8	6	3	19	328	64	34	45	45	76	224	3024
% share of total machining time	50.7	9.6	5.90	4.8	2.10	0.30	0.20	0.10	0.60	10.8	2.10	1.10	1.50	1.50	2.50	7.40	100%

Table 5.6: Interruptions in Different Classes of Work: Work Sampling Summary Form

UVW manufacturing company

Department: ____________

Date: ____________

Operation	Engineer	Supply	Quality	Mechanic	Supervision	Switch on lights	Miscellaneous	Actually working	No. of interferences	Total observations	Percent allowance
Bench	1	11	1	0	12	1	125	2850	26	3001	0.949
Bench machine	0	2	0	0	5	0	75	995	7	1077	
Machine	0	1	6	11	9	0	35	1200	27	1262	
Spray	0	0	0	7	45	0	265	1510	52	1827	

The program aims to maximize the value of % PACE expressed in the following formula. If it is observed that the value is already high, then the objective is to try and improve it. The observers are usually industrial engineers who are well versed in the operations of the department under scrutiny. A training and familiarization time of about 2 weeks is afforded to all observers to allow them to acclimatize themselves with the division being studied:

$$\%\ \text{PACE} = \frac{\{\text{Number assigned} \pm \text{loans} - (\text{idle and out of area})\} \times \text{effort rate} \times 100}{\text{Number assigned} \pm (\text{loans} - \text{out of area})}$$

The objective is to obtain a value of % PACE equal to 85%–100% efficiency. It is obvious that the efficiency cannot exceed 100%, but consider a value equal to 65%, for instance, would imply a waste of about 35% that is usually too excessive.

One of the advantages of this system is that it rapidly alerts supervisors of newly emerging "nonvalue adding" or uneconomic situations as far as use of the workforce is concerned.

5.9 EXAMPLES OF WORK SAMPLING

The use of work sampling is not limited to manufacturing environments alone. Work sampling is also used in nonmanufacturing environments such as banks, restaurants, hospitals, and department stores to determine daily and hourly personnel requirements and for cost control. Large companies usually conduct work-sampling studies of the activities of personnel in many departments to compute productivity.

Consider the following example:

EXAMPLE 5.1

Assuming an observation period of 300 h during which 27,000 parts have been processed. A total of 1600 workers worked while they were being observed and 400 workers were idle. The performance rating allocated was 80% and allowances were 10%. What should be the standard number of units processed per hour?

It will be observed that the total number of workers = 1600 + 400 = 2000.

Thus the proportion of busy workers = 1600/2000 = 80%.

Proportion of idle workers = 400/2000 = 20%.

Thus, the actual number of working hours out of 300 = 0.8 × 300 = 240 h.

$$\text{Normal time} = \frac{\text{Actual time} \times \text{Performance rating}}{100}$$
$$= \frac{(240 \times 80)}{100}$$
$$= 192\ \text{h}$$

Allowance = 10% = 0.1 × 300 h = 30 h.

Standard time = 192 + 30 = 222 h.

Thus, the standard number of units produced per hour = 27,000/222 = 122 units.

EXAMPLE 5.2

Consider the case of a pharmacy where the supervisor is trying to determine the staffing requirements.

The objective is to determine whether there are any improvements that could be made to the workload as well as the staffing patterns of another pharmacy that is located some distance away. Thus, the main objective in this case is to ascertain what percentage of an employee's day was devoted to a certain activity.

This is a case of computing the proportion of active time as a percentage of total time spent working. Total time is the sum of active time and idle time. Furthermore, the objective is to compute the percentage of active time spent in performing a specific activity. This can be easily accomplished by taking the ratio of time taken to perform the activity under consideration to the total time spent on the job.

In this particular case, the sample size was one pharmacy location and PMTS was used to compute the ratios. These can then be compared with the actual values. Similarly, time values and efficiency can be "visualized" in the case of a pharmacy location that currently does not exist but is likely impending.

The study was performed over 3 weeks including all shifts and a total of 2400 observations were recorded.

EXAMPLE 5.3

This is a case of a task analysis study performed at a 75-bed rural hospital with comprehensive pharmaceutical services. It is the activity log of a pharmacist's activities in this environment.

The objective is to determine the manner in which a pharmacist's time was utilized: namely to investigate what portion of the workday was spent in clinical pharmacy activities. Moreover, the amount of time allocated to the pharmacists in this specific task was sought to be compared with that of peers (other pharmacists) in a small hospital. As a secondary objective, it was also sought to determine the degree of usefulness of potential support staff.

During the course of this study, one pharmacy location was studied and samples were obtained over 5 min intervals that were taken to be a fixed time interval.

The study was conducted over 18 days, which were randomly selected over a 9-month period. Seven trained observers directly observed activities and a total of 2500 observations were recorded.

EXAMPLE 5.4

In this example, work sampling was used to assess nursing efficiency. The objective was to determine how nurses utilized their time. A total of 20 registered nurses, about 15 certified nursing assistants, and 5–7 secretaries were the object of observation.

The observations were performed for a duration of about 30 min/shift. The data was obtained through direct observation.

The study was conducted for 48 h comprising 8 day shifts, 8 evening shifts, and 4 night shifts on weekdays as well as on weekends. A total of 4000 observations were recorded.

5.10 SUMMARY

The concept of work sampling was addressed in detail in this chapter. The context in which work sampling is used as well as its advantages and potential disadvantages were discussed. Potential areas of work in which work sampling may be applied with a good deal of advantage were also mentioned. Additionally, several real life examples of the application of work sampling were presented.

From the preceding discussion, one can conclude that the techniques of work sampling can be used advantageously in a wide variety of situations and in many different scenarios, not just limited to manufacturing. The entire concept revolves around studying the efficiency of different operations in order to try and maximize it. Efficient operations lead to efficiently run organizations, which in turn are conducive to higher profits.

6 Predetermined Motion Time System (PMTS)

6.1 THE BASIC CONCEPT OF PMTS

Every industrial task is composed of a sequence of basic human activities that have to be performed optimally in order to achieve the desired results. Examples of such activities include: reach, move, turn, apply pressure, grasp, position, release, disengage, eye times (for observation), and body, leg, and foot motion.

Predetermined motion time systems (PMTSs) are methods of setting basic times for performing the aforementioned human activities that are necessary for carrying out a job or task. PMTSs are in essence work measurement systems that perform three main functions:

1. Analyze work
2. Break down the work into basic human movements. These are classified according to the nature of each movement and the conditions under which it is performed
3. Compute the time necessary to perform an operation. This is accomplished by adding the times required to perform individual activities

Thus, PMTS can be described as a motion-based system of measuring work. Each motion is assigned a specific code and a time value. This value is "predetermined." Adding the time values assigned to each constituent motion yields the total amount of time required to complete any given task. It will be appreciated that PMTS affords the engineer an objective standard for measuring work and, in turn, productivity.

The main advantages of PMTS are listed as follows:

- Higher degree of reliability
- Higher degree of consistency
- Higher degree of accuracy
- Can be used as a benchmarking tool
- Can be used to measure improvement objectively over time
- The use of PMTS eliminates the requirement performance rating
- PMTS enables calculation of time prior to starting production. This renders it useful in method design, equipment selection, and design along with production planning and control
- Time standard for a job can be computed without accessing the workplace
- The time and cost associated with finding the standard time for a job are considerably reduced

The three main types of PMTS that will be discussed in this chapter are

1. Method time measurement (MTM)
2. Work factor
3. Maynard operation sequence technique (MOST)

6.2 METHOD TIME MEASUREMENT

Method time measurement (MTM) is one of the PMTS developed by the MTM Association of Standards and Research. It was first released in 1948 and several variations are in existence today, including MTM1, MTM2, MTM-UAS, MTM-MEK, and MTM-B. For the sake of this chapter we will address the basic concept of MTM1 in detail.

MTM analyzes an industrial job, any manual operation or method into the basic motions or human movements required to perform it. It also assigns to each type of motion a predetermined time standard, which is determined by the nature of the motion and the conditions under which it is made. When all times for all motions are added up, the sum provides the normal time for the job. Standard time can be determined by adding suitable allowances. The most relevant industrial operations along with their representative symbols are described in the following list.

1. Reach (R): Reach is the basic element used when the predominant purpose of the motion is to move hands or fingers to a definite destination or to a general location.

   The time for making a reach varies with the following factors:

   a. Condition (nature of destination)
   b. Length of motion
   c. Type of reach (i.e., whether hands move/accelerate/decelerate at the beginning/end of reach or not)

   Classes of reach: There are five classes of reach. The time to perform a reach is affected by the nature of the object toward which reach is made. There are five cases for the motion reach.

   Case A reach: Reach to an object in other hand or to an object in fixed location or on which other hand rests.

   Case B reach: Reach to object whose general location is known. Location may vary a little from one cycle to another.

   Case C reach: Reach to object jumbled with other objects in a group. Search and select may be involved in this case.

   Case D reach: Reach to a very small object or where accurate group is needed.

   Case E reach: Reach to indefinite location to get hand into position for body balance or next move or out of way.

   True path is defined as the length of motion, not just the straight-line distance between the two terminal points. This is important because it accounts for real-world working conditions. The straight-line distance between two points is defined by the shortest distance between those points. This may not necessarily correspond to the actual nature of the reach.

   There are three types of reach to be taken into consideration:

   a. Hand is not moving at beginning and at end of reach
   b. Hand is moving at either beginning or at end of reach
   c. Hand is in motion at both beginning and at end of reach

2. Move (M): Move is the basic element used when the predominant purpose is to transport an object to a destination. There are three classes of moves.

   Case A move: Object to other hand or against stop

Case B move: Object to approximate or indefinite location

Case C move: Object to exact location

The time for move will depend on the following variables:

a. Length of motion. This is analyzed in a manner similar to reach
b. Condition (nature of destination) as governed by the aforementioned types of moves
c. Types of move that are the same as in the case of "reach"
d. Weight factor (static and dynamic)

The weight of the object has a dual effect on the move time. The amount of time necessitated to affect the move is increased. Similarly, a higher weight produces an impulse to initial hesitation prior to movement also characterized as a "pause." This further increases time.

3. Turn (T): Turn is the basic element used when the hand, either empty or loaded, is turned. Such movement rotates the hand, wrist, and forearm about the long axis of the forearm. The amount per length of turn is measured by the degrees by which the hand, wrist, and forearm are turned from their natural position. The table for turn contains time values (i.e., TMU) from 30° turn to 180° turn with increments of 15°.

   The time to affect a turn depends on two variables:

   a. Degrees turned
   b. Weight factor

   The turn time is dependent on the weight of the object being turned if hand is not empty.

4. Apply pressure (AP): Application of pressure implies the exertion of precise control. It can be visualized as a distinct pause before performing any subsequent actions. It provides full cycle time or development by the components as related to other motions. Pressure can be applied in two distinct ways as described below:

   Case A: Application of pressure only. This would naturally require less time than B.

   Case B: Regress or sequence and apply pressure. It needs precision and therefore takes more than normal time.

5. Grasp (G): "Grasp" is generally performed with the main purpose of securing sufficient control on one or many objects with fingers or with hands. It is done in order to allow the performance of the next basic element. It starts when the preceding basic element has ended and stops when the succeeding element has started. It includes the actions of "searching" and "selecting."

6. Position (P): Position is the basic element used to achieve alignment and engages one object with another object. The motions used in this context are very minor and thus cannot be classified as other basic elements. The time for position is affected by the following variables:

   a. *Class of fit*: This can be classified as either loose fit, close fit, or an exact fit
   b. *Symmetry*: Symmetrical (S), nonsymmetrical (NS) semisymmetrical (SS)
   c. *Ease of handling*: Whether easy to handle or difficult to handle

7. Release (RL): Release is the term used to describe the basic element used to relinquish control of an object (to release and let go of an object that was being held by the hand)

   The "release" function can be classified into two categories:

   a. *Normal release*: During normal release, the fingers are opened to let go of the object

   b. *Contact release*: Contact release occurs and is completed when the following reach motion starts

8. Disengage (D): Disengage is the basic element used to break contact between one object and another. The objects are separated from one another.

   Disengage times depend on

   a. Class of fit

   b. Ease of handling

   c. Care in handling

9. Eye times (ET: eye travel and EF: eye focus): Most industrial and work activities involve time required for moving. Focusing the eye is not a limiting factor and therefore it does not affect the time required to perform the operation. Yet, there are some instances, especially in precision work that involve extensive amount of time required for eye travel and focus in order to perform an activity. These two types of times are described in MTM as ET time and EF time.

   EF time can be defined as the time needed to focus the eyes on an object and look at it long enough to identify certain readily distinguishable characteristics within the area. This may be seen without shifting the eyes. The distance between points to and from which the eye travels and the perpendicular distance from the eye to the line of travel affect ET time.

10. Body, leg, and foot movement: Foot motions involve motions where the foot is moved with the ankle used as a hinge or the instep used as the fulcrum of the motion. This is usually accomplished in the vertical direction.

    Leg motions involve motions of the foreleg or the entire leg. The knee or hip serves as a pivot. For the sake of time measurement, there is practically no difference between foreleg movement and entire leg movement. Thus, both motions are considered identical from this viewpoint.

    Sidestep motions involve movements wherein the body is displaced sideways in order to access a separate work area. A variation of this body movement is the "turn body" movement. It occurs when a worker turns the body to a new location while stepping from a workstation.

    Bend is the term used for a body movement wherein the body is bent at the waist with the upper part of the torso lowered to bring the hands to a position that would afford easy access. Such access would not be possible if the worker maintained an erect posture.

    Stoop is the term used to describe a body movement wherein the body is lowered by bending the knees. The stooping ends when the hands access the object that is situated close to the floor.

    Sitting occurs when the body is lowered so as to position itself on a chair or a bench. This action ends when the body is seated. Standing begins with the feet on the floor and ends when the body has assumed an erect position.

Tables 6.1 through 6.9 depict the MTM values associated with each basic activity. Table 6.1 is an encapsulation of the system and provides an easy to use reference for the time measurement analysts "on the go." Tables 6.2 through 6.9 should be used in order to gain an in-depth understanding of the system as well as to build accurate standards. A time measurement unit (TMU) is the equivalent of 0.036 s. Thus, the amount of time required to perform an operation is computed by determining the total value of TMUs necessary to perform an operation and converting that value to a time value mainly expressed in terms of number of seconds.

6.3 PROCEDURE TO APPLY MTM

The following section describes the steps to be undertaken in order to apply the MTM system to a work situation:

1. *Establishment of the basic method*: A method may either be visualized or observed. It can only be observed if it currently exists. On the other hand, a method that does not exist but its potential future application is being contemplated can be visualized. This stage involves a broad description of the method under consideration, its purpose and general details.

Table 6.1: Application of MTM Values to Basic Motions

Hand and Arm Motions			Body, Leg, and Eye Motions	
Reach or Move	**TMU**			**TMU**
1″	2		Simple foot motion	10
2″	4		Foot motion with pressure	20
3″ to 12″	4 + length of motion		Leg motion	10
Over 12″	3 + length of motion		Side step case 1	20
For type 2 reaches and moves, use length of motions only			Side step case 2	40
Position			Eye time	10
Fit	**Symmetrical**	**Other**	Bend, stoop, or kneel on one knee	35
Loose	10	15	Arise	35
Close	20	25	Kneel on both knees	80
Exact	50	55	Arise	90
Turn-Apply Pressure			Sit	40
Turn		6	Stand	50
Apply pressure		20	Walk per pace	17
Grasp			1 TMU = 0.00001 h	
Simple		2	1 TMU = 0.0006 min	
Regrasp or transfer		6	1 TMU = 0.036 s	
Complex		10		
Disengage				
Loose		5		
Close		10		
Exact		30		

Table 6.2: Individual Values for the "Reach" Function

Distance Moved in Inches	Time (TMU)				Hand in Motion		Description
	A	B	C or D	E	A	B	
0.75 or less	2	2	2	2	1.6	1.6	A: Reach to an object in fixed location or to object in another hand or on which other hand rests
1	2.5	2.5	3.6	2.4	2.3	2.3	
2	4	4	5.9	3.8	3.8	2.7	
3	5.3	5.3	7.3	5.3	4.5	3.6	
4	6.1	6.4	8.4	6.8	4.9	4.3	B: Reach to a single object in location which may vary slightly from one cycle to another
5	6.5	7.8	9.4	7.4	5.3	5	
6	7	8.6	10.1	8	5.7	5.7	
7	7.4	9.3	10.8	8.7	6.1	6.5	C: Reach to an object jumbled with other objects in a group so that search and select occur
8	7.9	10.1	11.5	9.3	6.5	7.2	
9	8.3	10.8	12.2	9.9	6.9	7.9	
10	8.7	11.5	12.9	10.5	7.3	8.6	D: Reach to a very small object or where accurate grasp is required
12	9.6	12.9	14.2	11.8	8.1	10.1	
14	10.5	14.4	15.6	13	8.9	11.5	
16	11.4	15.8	17	14.2	9.7	12.9	E: Reach to an indefinite location to get hand in position for body balance or next motion or out of way
18	12.3	17.2	18.4	15.5	10.5	14.4	
20	13.1	18.6	19.8	16.7	11.3	15.8	
22	14	20.1	21.2	18	12.1	17.3	
24	14.9	21.5	22.5	19.2	12.9	18.8	
26	15.8	22.9	23.9	20.4	13.7	20.2	
28	16.7	24.4	25.3	21.7	14.5	21.7	
30	17.5	25.8	26.7	22.9	15.3	23.2	

2. *Organization of information*: Information can be classified as tangible and intangible. This step involves collection of all information that is tangible in nature. The following are examples of tangible information in the context of MTM application:

   - Production requirements
   - Equipment and tools
   - Materials and parts
   - Location and conditions
   - Quality requirements
   - Relevant information that can be observed from similar operations

   Access to the aforementioned information is then reported on the back of an MTM analysis sheet. It should cover pertinent information related to operation (clear and concise description of the operation and the operation sequence), location (location description should be clear enough to enable an observer to precisely locate the spot, as well machine numbers and any other relevant information needs to be clearly reported), parts (information related to part number, name, etc. along with a drawing of the part), materials

Table 6.3: Individual Values for the "Move" Function

Distance Moved in Inches	Time (TMU) A	B	C	Hand in Motion (B)	Weight Allowance Weight (lbs) Up to	Factor	Constant TMU	Description
0.75 or less	2	2	2	1.7	2.5	1.0	0	A: Move object to other hand or against stop
1	2.5	2.9	3.4	2.3				
2	3.6	4.6	5.2	2.9	7.5	1.06	2.2	B: Move object to approximate or indefinite location
3	4.9	5.7	6.7	3.6				
4	6.1	6.9	8	4.3	12.5	1.11	3.9	C: Move object to exact location
5	7.3	8	9.2	5				
6	8.1	8.9	10.3	5.7	17.5	1.17	5.6	
7	8.9	9.7	11.1	6.5				
8	9.7	10.6	11.8	7.2	22.5	1.22	7.4	
9	10.5	11.5	12.7	7.9				
10	11.3	12.2	13.5	8.6				
12	12.9	13.4	15.2	10	27.5	1.28	9.1	
14	14.4	14.6	16.9	11.4				
16	16	15.8	18.7	12.8	32.5	1.33	10.8	
18	17.6	17	20.4	14.2				
20	19.2	18.2	22.1	15.6	37.5	1.39	12.5	
22	20.8	19.4	23.8	17				
24	22.4	20.6	25.5	18.4	42.5	1.44	14.3	
26	24	21.8	27.3	19.8				
28	25.5	23.1	29	21.2	47.5	1.5	16	
30	27.1	24.3	30.7	22.7				

Table 6.4: Individual Values for "Turn and Apply Pressure" Function

	Time (TMU) for Degrees Turned										
Weight	30	45	60	75	90	105	120	135	150	165	180
0–2 pounds	2.8	3.5	4.1	4.8	5.4	6.1	6.8	7.4	8.1	8.7	9.4
2.1–10 pounds	4.4	5.5	6.5	7.5	8.5	9.6	10.6	11.6	12.7	13.7	14.8
10.1–35 pounds	8.4	10.5	12.3	14.4	16.2	18.3	20.4	22.2	24.3	26.1	28.2

Apply pressure case 1: 16.2 TMU

Apply pressure case 2: 10.6 TMU

Table 6.5: Individual Values for the "Grasp" Function

Case	Time (TMU)	Description
1A	2.0	Pick up grasp-small, medium, or large object by itself, easily grasped
1B	3.5	Very small object or object lying close against a flat surface
1C1	7.3	Interference with grasp on bottom and one side of nearly cylindrical object. Diameter larger than 0.5 in.
1C2	8.7	Interference with grasp on bottom and one side of nearly cylindrical object. Diameter 0.25–0.5 in.
1C3	10.8	Interference with grasp on bottom and one side of nearly cylindrical object. Diameter less than 0.25 in.
2	5.6	Regrasp
3	5.6	Transfer grasp
4A	7.3	Object jumbled with other objects. Search and select occurs. Larger than 1″ × 1″ × 1″
4B	9.1	Object jumbled with other objects. Search and select occurs. Larger than 1″ × 1″ × 1″
4C	12.9	Object jumbled with other objects. Search and select occurs. Smaller than 0.24″ × 0.25″ × 0.125″
5	0	Contact, sliding, or hook grasp

Table 6.6: Individual Values for the "Position" Function

Class of Fit		Symmetry	Easy to Handle	Difficult to Handle
1: Loose	No pressure required	S	5.6	11.2
		SS	9.1	14.7
		NS	10.4	16.0
2: Close	Light pressure required	S	16.2	21.8
		SS	19.7	25.3
		NS	21.0	26.6
3: Exact	Heavy pressure required	S	43.0	48.6
		SS	46.5	52.1
		NS	47.8	53.4

Table 6.7: Individual Values for "Release" Function

Case	Time (TMU)	Description
1	2.0	Normal release performed by opening fingers as an independent motion
2	0	Contact release

(physical characteristics of the materials such as weight, color, size, and flexibility), equipment (equipment needs to be identified by name, numbers, and drawings, which includes auxiliary devices), quality requirements (factors such as tolerance, surface finish, and material handling), and tool and part sketches (freehand sketches of tools and parts, especially ones that help to determine the types of motions to be used).

Table 6.8: Individual Values for "Disengage" Function

Class of Fit	Easy to Handle	Difficult to Handle
1: Loose: Very slight effort, blends with subsequent move	4.0	5.7
2: Close: Normal effort, slight recoil	7.5	11.8
3: Tight: Considerable effort, hand recoils markedly	22.9	34.7

Table 6.9: Individual Values for "Body, Leg, and Foot Motions" Function

Description	Symbol	Distance	Time (TMU)
Foot motion hinged at ankle	FM	Up to 4″	8.5
With heavy pressure	FMP		19.1
Leg or foreleg motion	LM	Up to 6″	7.1
		Each additional inch	1.2
Sidestep:			Use reach or move time
Case 1: Complete when leading leg contacts floor	SS-C1	Less than 12″	
		12″	17.0
		Each additional inch	0.6
Case 2: Lagging leg must contact floor before next motion can be made	SS-C2	12″	34.1
		Each additional inch	1.1
Bend, stoop, or kneel on one knee	B, S, KOK		29.0
Arise	AB, AS, AK, OK		31.9
Kneel on floor			
Both knees	KBK		69.4
Arise	AKBK		76.7
Sit	SIT		34.7
Stand from sitting position	STD		43.4
Turn body 45°–90°			
Case 1: Complete when leading leg contacts floor	TBC 1		18.6
Case 2: Lagging leg must contact floor before next motion can be made	TBC 2		37.2
Walk	W-FT	Per foot	5.3
Walk	W-P	Per pace	15.0

3. *Elemental breakdown*: The task needs to be broken down into individual elements that are described in sufficient detail similar to that commonly observed in a detailed time study. This includes an accurate representation of the workplace layout. The normal work simplification process (methods development) needs to be followed. Elemental sequence needs to be examined for accuracy. Elements need to be checked for overlap, especially ones that occur between manual work elements and machine elements. The elements also need to be checked to ensure that the rules of motion economy are being adhered to.

4. *Conduct MTM analysis*: The steps outlined herein are to be followed when conducting the actual MTM analysis:

   a. Determine a detailed elemental breakdown by

   i. Separating fixed and variable elements
   ii. Shortening each element as much as possible
   iii. As far as possible assigning no more than 10 individual motions by each hand to an element
   iv. Describing all elements in the right sequence

b. Each element needs to be described in detail and its description needs to be recorded at the top of an MTM analysis sheet

c. Record on the MTM analysis sheet motions performed by the left hand as well as the right hand as follows:
   i. All motions need to be recorded
   ii. Highlight nonlimiting simultaneous motions and strike out nonlimiting combined motions
   iii. All motions performed by the foot, leg, and the body under the column for "right hand"
   iv. Only one element needs to be analyzed and completed at one time
   v. The description of each motion needs to be complete and logical enough to warrant award of MTM motion time

d. The motion sequence needs to be checked for errors in visualization, observation, and recording

e. The final study should be signed and dated for reasons for traceability

5. *Completion of the study*: The study is completed by undertaking the following steps:
   a. The elements need to be identified and described in sequence and the assignment of TMUs needs to be done accordingly
   b. The TMUs are then converted to equivalent time values by using the appropriate conversion factor
   c. Issues such as personal needs, unavoidable delays, fatigue etc. are compensated for by using the allowance factor
   d. The allowed element time is determined and recorded as "time + allowance"
   e. The frequency of occurrence of each element is recorded as occurrence per cycle
   f. The allowed time per element per cycle is recorded as the total time required

The study should always be carefully checked for mathematical errors and any errors that may accrue from improper application of the procedure.

Table 6.10 depicts several conventions that are used to identify basic motions in a typical industrial setting. Note that some of these motions can be used in an office setting as well. Thus, it can be pointed out that basic MTM principles are universal in their appeal and application.

Table 6.11 presents a case study in the form of a simple industrial operation. Note the manner in which MTM data is utilized in order to compute the amount of time required to perform an industrial operation. The table describes a simple

Table 6.10: Conventions for Identifying Basic Motions

Motion	Symbol	Motion Detail	Convention
Reach	R	Reach, 5 in., case A, hand in motion at start	mR5A
Move	M	Move, 9 in., case B, Weight 15 pounds Move, 15 in., case A, hand in motion at end	M9B15 M15Am
Turn	T	Turn, 45° arc, small load	T45S
Apply pressure	AP	Apply pressure, case 2	AP2
Grasp	G	Grasp, case 1B	G1B
Release	RL	Contact release	RL2
Position	P	Position, close fit, nonsymmetrical fit, part easy to handle	P2NSE
Disengage	D	Disengage, close fit, part easy to handle	D2E

Table 6.11: Computing the Time Required for an Assembly Operation Using MTM

	Job Description				
	Assemble Transformer to Base Plate	Analyst			
Element	Description	Left Hand	TMU	Right Hand	Description
1	Move hand to washer	R14C	15.6	R14B	R14C: Reach 14 in. to an object jumbled with other objects. Class C reach R14B: Reach 14 in. to single object. Class B reach
2	Grasp first washer	G4B	9.1	G1A	Grasp transformer
3	Move hand clear of container	M2B			Hold in a box
4	Palm washer	G2	5.6		G2: Regrasp, case 2 grasp
5	To second washer	R2C	5.9		
6	Grasp washer	G4B	9.1	M14C	Transformer to plate
7	Move washers to area	M10B	16.9	M14C	

sequence of operations involving the assembly process of a transformer to base plate. The total time required to perform the operation is computed by adding the TMUs for individual operations and converting them to actual time.

6.4 WORK FACTOR

The main purpose of a PMTS is to provide a useful time measurement tool. Such a tool will enable the user to quantitatively measure the amount of manual and mental work performed by an operator with as little error (as much accuracy) as possible. The work factor system aims to accomplish this objective. There are six different work factor systems. They are described as follows.

6.4.1 Detailed Work Factor

Detailed work factor (DWF) is used for functions that need accurate time standards. It is used in work involving a high amount of repetition as well as mass production work. It is also used in work involving short cycles. The savings resulting from the use of accurate time standards are often observed to be greater than the expense necessary to establish these standards. DWF is extensively used in many factory and office work situations since a large amount of such work involves the aforementioned conditions. This is a general basic level system. Times required for elemental movements are achieved by determining the work factor involved.

According to the work factor system, there are four major variables that affect the time required to perform manual motions. These variables are listed below:

- Body member used
- Distance moved measured in inches
- Manual control required, measured in work factors, defined or dimensional
- Weight or resistance involved, measured in pounds, converted to work factor

1. *Body member used:* There are six definite body members involved and motion times for each member are provided: Finger or hand, arm, forearm, swivel, trunk, foot, and leg. Notations used for each member are as follows: Finger or hand (FH), arm (A), forearm swivel (S), trunk (T), foot (FT), and leg (L). In addition to the aforementioned body members, DWF tables also provide values for walking time.
2. *Distance:* All distances (except those with change in direction) are measurement in straight lines between the starting and stopping points of the motion as described by the body members. Tables specify the point at which various body members' motion has to be measured. If the change of direction is intentionally done during movement it will be taken care of by a work factor. This is important since the distance being measured is in a straight line.
3. *Manual control:* Various types of motions occasionally require varying degrees of manual control. This is expected to increase the motion time. The manual control necessary is specified by the work factor. The level of difficulty involved is illustrated by the following classification of the types and degrees of control: definite stop work factor (D), directional control work factor (steer) (s), care work factor (precaution) (p), and change of direction work factor (U).
4. *Weight or resistance:* The effect of weight on the time required to perform a motion is due to the weight of the object being handled as well as the need to exert force in order to overcome resistance. The effect of weight on time is dependent on the body member used and on the sex of the operator. The two variables, distance and body member, are indicated in inches and the body member used, respectively. They are independent of and are not affected by work factors.

In simplistic terms, work factor can be considered as a means of describing the motion according to the amount of weight or resistance involved in its performance. Weight is taken as a work factor in order to compute motion times in work factor system.

The following sequence of steps is involved in using the table:

1. Analyze the motion whose time is to be determined from the table and compute the following information:
   a. Spot the body member involved
   b. Compute the amount of distance moved
2. On the basis of this information, identify the particular row of the table.
3. The first entry of the row under the heading basic provides the time of this motion without any work factor of manual control and weight/resistance involved.
4. Next, the motion is analyzed to determine the control work factors involved. As is evident from the table, there can be at most four work factors and four time values which are indicated corresponding to 1, 2, 3, and 4 work factors, respectively.
5. Finally, the weight work factors for men and women are given at the bottom of each table. These are used for determining the weight work factor.

Table 6.12 depicts the data that is used to compute operation time using the work factor. It is to be borne in mind that a time unit is the equivalent of 0.006 s. Thus, the amount of time required to perform an operation is computed by determining the total value of time units necessary to perform an operation and converting that value to a time value mainly expressed in terms of number of seconds.

This is followed by Table 6.13 which demonstrates the system in practice.

A complex grasp operation is analyzed using the work factor in Table 6.13. Consider an operation involving grasping two small spring clips, 1/16 × 1/16 in., two at a time. The grasping is accomplished one in each hand simultaneously and the springs have a tendency to nest.

6.4.2 Simplified Work Factor

This is a general higher level system and is suitable for medium quantity production developed from a DWF system. It is quite effective in preparing estimates before actual production starts. Like other higher level systems, it is easier and is rapidly used. Time units used in work factor table are in ten thousandths of a minute.

6.4.3 Abbreviated Work Factor

This is a higher level system developed from the DWF system. It is especially suitable for small quantity job shops. These systems were developed to fill the need for a very simple system of predetermined time standards. Their greatest advantage is that they provide a rapid measurement technique. It is unnecessary to refer to a separate table of time values to use this system. The abbreviated time unit is 0.005 (recorded as 5) minute rather than 0.0001 min used in the Detailed and Mento systems. The accuracy of the abbreviated systems is expected to average +12% of the detailed system if it is applied correctly to the appropriate type of work.

6.4.4 Ready Work Factor

Ready work factor (RWF) is a higher level modified work factor system and is meant to be used by operators who are not familiar with the details of time study techniques. It is normally utilized in evaluating manual work in industry and other similar places. This technique has been developed to satisfy the requirement for a simple method of evaluating, measuring, and comparing the time required for manual motions.

Table 6.12: Work Factor Motion Time Table: Time Measured in TMUs

Distance Moved (in.)	Basic	Work Factors 1	2	3	4	Distance Moved (in.)	Basic	Work Factors 1	2	3	4
(A) Arm measured at knuckles						**(L) Leg measured at ankle**					
1	18	26	34	40	46	1	21	30	39	46	53
2	20	29	37	44	50	2	23	33	42	51	58
3	22	32	41	50	57	3	26	37	48	57	65
4	26	38	48	58	66	4	30	43	55	66	76
5	29	43	55	65	75	5	34	49	63	75	86
6	32	47	60	72	83	6	37	54	69	83	95
7	35	51	65	78	90	7	40	59	75	90	103
8	38	54	70	84	96	8	43	63	80	96	110
9	40	58	74	89	102	9	46	66	85	102	117
10	42	61	78	93	107	10	48	70	89	107	123
11	44	63	81	98	112	11	50	72	94	112	129
12	46	65	85	102	117	12	52	75	97	117	134
13	47	67	88	105	121	13	54	77	101	121	139
14	49	69	90	109	125	14	56	80	103	125	144
15	51	71	92	113	129	15	58	82	106	130	149
16	52	73	94	115	133	16	60	84	108	133	153
17	54	75	96	118	137	17	62	86	111	135	158
18	55	76	98	120	140	18	63	88	113	137	161
19	56	78	100	122	142	19	65	90	115	140	164
20	58	80	102	124	144	20	67	92	117	142	166
22	61	83	106	128	148	22	70	96	121	147	171
24	63	86	109	131	152	24	73	99	126	151	175
26	66	90	113	135	156	26	75	103	130	155	179
28	68	93	116	139	159	28	78	107	134	159	183
30	70	96	119	142	163	30	81	110	37	163	187
35	76	103	128	151	171	35	87	118	147	173	197
40	81	109	135	159	179	40	93	126	155	182	206
Weight (pounds)						Weight (pounds)					
Male	2	7	13	20	UP	Male	8	42	UP		
Female	1	3.5	6.5	10	UP	Female	4	21	UP		
(T) Trunk measured at shoulder						**(FH) Finger or hand measured at finger tip**					
1	26	38	49	58	67	1	16	23	29	35	40
2	29	42	53	64	73	2	17	25	32	38	44
3	32	47	60	72	82	3	19	28	36	43	49
4	38	55	70	84	96	4	23	33	42	50	58
5	43	62	79	95	109	Weight (pounds)					

(Continued)

Table 6.12: (*Continued*) Work Factor Motion Time Table: Time Measured in TMUs

Distance Moved (in.)	Basic	Work Factors 1	2	3	4
6	47	68	87	105	120
7	51	74	95	114	130
8	54	79	101	121	139
9	58	84	107	128	147
10	61	88	113	135	155
11	63	91	118	141	162
12	66	94	123	147	169
13	68	97	127	153	175
14	71	100	130	158	182
15	73	103	133	163	188
16	75	105	136	167	193
17	78	108	139	170	199
18	80	111	142	173	203
19	82	113	145	176	206
20	84	116	148	179	209
Weight (pounds)					
Male	11	58	UP		
Female	5.5	29	UP		

Distance Moved (in.)	Basic	Work Factors 1	2	3	4
Male	0.67	2.5	4	UP	
Female	0.33	1.25	2	UP	
(FT) Foot measured at toe					
1	20	29	37	44	51
2	22	32	40	48	55
3	24	35	45	55	63
4	29	41	53	64	73
Weight (pounds)					
Male	5	22	UP		
Female	2.5	11	UP		
(FS) Forearm swivel measured at knuckles					
45°	17	22	28	32	37
90°	23	30	37	43	49
135°	28	36	44	52	58
180°	31	40	49	57	65
Torque (pound-in.)					
Male	3	13	UP		
Female	1.5	6.5	UP		

Work factor symbols	
W	Weight or resistance
S	Directional control (steer)
P	Care (precaution)
U	Change of direction
D	Definite stop

Table 6.13: Application of Work Factor to Determine Operation Time for a Simple Grasping Operation

Description	Motion Analysis	Work Factor (Time Units)
First grasp	F1	16
Second grasp	Negligible	0
Separate	V5F1P	115
Allowance for nested	V3/4F1P	17
Simo factor	50% of V53/4 F1P	66
Total grasp time		214 = 0.0214 min

Product designers, cost estimators, methods and tool engineers, office supervisors, and others familiar with production activities can be taught RWF. It is a very simple system which is particularly useful in measuring medium to long run operations with cycles of 0.15 min and longer. In the application of RWF, a simple set of preestablished time standards are used according to rules that are easy to understand, remember, and apply.

6.4.5 Brief Work Factor

Brief work factor (BWF) is relatively new and like the abbreviated systems uses a time unit of 0.005 min (recorded as 5). It was developed to make the format conform to the detailed system and the ready system.

6.4.6 Detailed Mento Factor

It is a basic level system which is used for measuring mental processes such as inspection, proofreading, color matching, and calculations and virtually any repetitive or semirepetitive and predominantly mental operation.

6.5 MAYNARD OPERATION SEQUENCE TECHNIQUE

Often, PMTS analyze work using the concept of basic motions. For instance, MTM1 used the 19 fundamental basic motions and developed time standards. This arrangement accounted for the spatial and sensory characteristics and other variables. The process was time consuming and resource intensive. In fact, it required a database of 300 time standards.

The complexity of analysis requires a high degree of skill on the part of the analyst. Nevertheless, the process of analyzing work and computing the operation time is not only time consuming but is often inconsistent as well. Recognizing the previously mentioned limitations of the traditional PMTS, a significant amount of work was undertaken in order to design more efficient means of application that drew upon the original data.

The Maynard operation sequence technique (MOST) is different from other PMTSs. Unlike MTM and work factor, for instance, it analyzes the sequence of activities instead of the basic motions necessary to perform an operation. The sequence of activities is predefined and remains constant throughout the operation. It is imperative that the analysts apply the sequence model with no alterations and every factor needs to be considered logically and consistently. MOST has been shown to be more consistent than other PMTS.

The main advantages of MOST are listed as follows:

- Significant cost reduction in view of simplified procedures
- Simplified procedures lead to reduced paperwork
- Simplified procedures and reduced paperwork imply less input for the same output. Thus MOST results in higher productivity
- The main objective of MOST is to streamline operations by the application of PMTS principles
- MOST systems are easy to learn and use
- MOST is very effective in quickly pinpointing inefficient operations
- Results obtained from MOST systems are highly consistent
- Results from MOST are highly accurate (5% within a 95% confidence interval)
- MOST can be applied effectively to any methodical work system

MOST computes the normal time to perform an operation/process by adhering to the following steps:

1. Deconstruct the operation/process into individual steps/units
2. Analyze the motions in each step/unit using a standard MOST method sequence
3. Assign indices to the parameters constituting the method sequence for each task
4. Add the values of individual indices to arrive at a time value for each step/unit
5. Add the time values for all the steps/units to compute the normal time to perform the operation/process

The position of the MOST system in the work hierarchy is as represented in Figure 6.1.

6.5.1 Basic MOST

The concept of the Basic MOST is that it focuses on the movement of objects. It will be appreciated that most industrial tasks involve movement of objects such as tools, parts, and miscellaneous objects from one place to another. Most assembly, disassembly, and maintenance operations involve moving objects and manipulating tools by moving them from one location to another.

The collection of basic motion elements is referred to as motion aggregates. This is the building block of the MOST system. The motion aggregates focus on moving objects around and are referred to as activity sequence models in MOST terminology.

There are three activity sequence models in the MOST system. Each sequence is composed of a standard sequence of actions. The three basic activity sequence models are listed as follows:

1. *General move*: This sequence model involves the free movement of an object through space from one point to another. For instance, picking up a component from the table and placing it on a fixture can be termed as a general move.

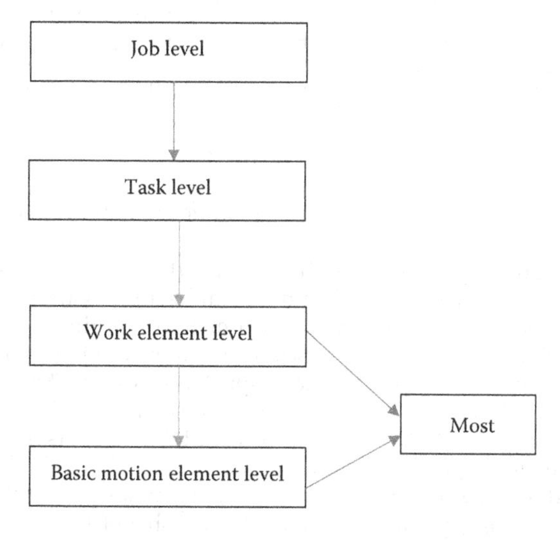

Figure 6.1 Position of MOST in the hierarchy of work activity.

2. *Controlled move*: This sequence model involves the movement of an object while it maintains contact with a reference surface. For instance, sliding a part along a table in order to affect an assembly is an example of controlled motion. Another example of controlled motion when an object is attached to another object and constrained in some manner during motion. For instance, operating a lever on a machine could be termed as an example of a controlled motion.
3. *Tool use*: This sequence model comes into play whenever an operation requires the use of a hand tool such as a hammer or a wrench.

Each activity sequence model is composed of a sequence of individual actions. Each action is referred to as a "sequence model parameter." Thus, each sequence model is composed of several parameters arranged in a particular sequence. The sequence parameters in the Basic MOST system are similar to basic motion elements in the MTM system.

The sequence model "general move" is composed of a sequence of basic parameters labeled A, B, G, and P. Each parameter is described as follows.

6.5.2 General Move

A: Action distance: Action distance is usually measured in the horizontal direction. It is used to describe the movement of fingers, hands, and feet. The movement may be performed either loaded or unloaded.

B: Body motion: This is usually measured vertically. Vertical body motions and actions in the vertical plane are measured by this parameter. Sitting and standing up could be cited as examples of such motions.

G: Gain control: A large number of industrial operations can only be performed after gaining control of an object or a tool. Such actions involve manipulation of the object, grasping of an object, etc. Therefore, some role is played by the fingers, hands, or feet in order to gain physical control of an object. This parameter is quite similar to the "grasp" element in the MTM system.

P: Placement: This parameter is used to describe the action involved to "place an object." This can include a variety of actions such as position, orient, place aside, or align an object once it has been moved to a specific location.

The general move is composed of the aforementioned parameters in a particular sequence:

$$A-B-G-A-B-P-A$$

It will be appreciated that the first three parameters in the sequence (ABG) describe the actions necessary to obtain an object.

The following three parameters (ABP) describe the actions necessary to move the object to the desired location. The final parameter (A) deals with the final motion which, in most cases, involves returning to the original position.

Each parameter is assigned a numerical value in the form of an index number that represents the time necessary to accomplish that action. The value of this number depends on the type of action, the motion content of the action, and the conditions under which the action is performed. Table 6.14 lists the actions and the values of the corresponding index numbers.

Once the index values for all parameters have been tabulated, they are added together and multiplied by 10 to determine the number of TMUs. For instance, consider the example of an activity that involves a worker walking five steps,

Table 6.14: MOST Parameters and Index Values for the General Move Activity Sequence Model

General Move Sequence Model: A-B-G-A-B-P-A

Index	A = Action Distance	B = Body Motion	G = Gain Control	P = Placement
0	Close ≤2 in.			Hold, toss
1	Within reach but >2 in.		Grasp light object with one or two hands	Lay aside Loose fit
3	1 or 2 steps	Bend and arise with 50% occurrence	Grasp object that is heavy or obstructed or hidden or interlocked	Adjustments, light pressure, double placement
6	3 or 4 steps	Bend and arise with 100% occurrence		Position with care, or precision of blinf, or obstructed or heavy pressure
10	5,6, or 7 steps	Sit or stand		
16	8,9, or 10 steps	Through door or climb on or off or stand and bend or bend and sit		

picking up a part from the floor, returning to his original position, and placing the part on the table.

In this situation, using the data from Table 6.14, the sequence of actions could be listed as follows:

$$A_{10} - B_6 - G_1 - A_{10} - B_0 - P_1 - A$$

A_{10} = Walking five steps
B_6 = Bend and rise
G_1 = Gain control of small part
A_{10} = Walk back to original position
B_0 = No body motion
P_1 = Place the part on the table
A = No motion

The sum of the index value is 28. Multiplying this value by 10 yields 280 TMUs which correspond to about 10 s.

6.5.3 Controlled Move

This sequence is used to address activities that involve operating a crank or a lever, a switch or any component that is attached to a machine. It also covers any motions that involve sliding an object over a surface. In the case of a controlled move, there are some parameters that need to be included in addition to the parameters listed for the general move. The sequence models for a controlled move includes the following additional activities:

M (Move controlled) XI (Pr ocess time) Align

For instance consider an activity involving engaging the feed level of a milling machine. The sequence models for such activity can be indexed as

Get	Actuate	Return
$A_1 - B_0 - G_1$	$M_1 - X_{10} - I_0$	A_0

A_1 = Reach to a lever a distance within reach
B_0 = No body motion
G_1 = Obtain a hold of the lever
M_1 = Move lever up to 12 in. to engage feed
X_{10} = Process time of about 3.5 s
I_0 = No alignment
A_0 = No return

6.5.4 Tool Use

The use of hand tools for effecting manual activities, such as fastening, loosening, cleaning, and cutting, is addressed by the tool use sequence model. Similarly activities that require some mental thought processes are also included in this sequence model. This model is a combination of the general move and controlled move activities.

Take for instance an activity that involves inspection operation. This activity can be described using the following sequence: Pick up a part, inspect two points on the part, and put it back on the conveyor. Thus we have the following sequence:

$$A_1 \quad B_0 \quad G_1 \quad A_1 \quad B_0 \quad P_0 \quad T_3 \quad A_1 \quad B_0 \quad P_1 \quad A_0$$

A_1 = Reach to part within reach
B_0 = No body motion
G_1 = Gain control of part
A_1 = Bring part within reach
B_0 = No body motion
P_0 = No placement
T_3 = Inspect two points
A_1 = Move part within reach
B_0 = No body motion
P_1 = Put part on the conveyor
A_0 = No return

The following section describes the different types of MOST systems.

6.5.5 Types of MOST Systems

There are two main types of MOST systems (in addition to Basic MOST) depending on the level of accuracy. These are described in some detail as follows.

6.5.5.1 Mini MOST

This comprises the lowest level of the MOST system. The level of detail and precision addressed at this level enables analysis of an operation that requires about 1500 repetitions. The operation may last anywhere from 2–10 s. In contrast to the Basic MOST, the index value total for a second is multiplied by 1 and converted to minutes or seconds. Typical applications include light press operations and manufacturing of printed circuit boards.

6.5.5.2 Maxi MOST

This comprises the highest level of the MOST system. It is used to analyze operations that are likely to be performed 150 times per week. The typical duration of such an operation may range from as little as 2 min to as much as several hours.

The index ranges for Maxi MOST accommodate the high amount of variations from one cycle to the next. At this level, the method descriptions resulting from Maxi MOST are highly practical for purposes of instruction. The index values used by Maxi MOST are identical to those used by Basic MOST. A multiplier of 100 is used in this case to obtain the TMU value. This value can then be converted into minutes or seconds as desired. Typical applications include job shop manufacturing such as ship building fabrication of an aircraft, maintenance operations, etc.

6.6 SUMMARY

This chapter presents the basic concept of predetermined time systems also abbreviated as PMTS. It deals with three main types of PMTS namely MTM, work factor, and MOST. The advantages of using PMTS are clearly stated at the beginning of the chapter. The main idea of using PMTS is to do away with the subjectivity and operator skill that is a major requirement of the stopwatch time study. Thus, PMTS can be effectively used by almost anyone as long as they are sufficiently skilled in method analysis. Besides, it is worth noting that most PMTSs enable the operator to "predetermine" the amount of time a particular operation should take assuming that all relevant parameters pertaining to that operation are known. Thus, the inclusion of the word "predetermined" in the vernacular.

7 Standard Data

7.1 THE CONCEPT OF STANDARD DATA

Standard data can be compared to a data set that comprises a collection of time values. This data set uses work elements from time studies or other sources of work measurement. This makes it unnecessary to restudy the work elements that have already been adequately timed in the past. Such elemental times are first extracted from the studies and then they are applied to tasks with similar or same elements. Standard data can be developed using graphs, tables, charts, formulas, and spreadsheet programs.

The greatest disadvantage of using time study and developing MTM standards is that the application of the principle requires minute observation of each motion necessary to perform each operation. Therefore, it is only suitable in the case of short and highly repetitive operations. The time required to study and analyze each operation can easily be equal to several hundred times the amount of time required to actually perform the operation itself. It is highly time consuming and expensive to develop a time standard.

This is one of the most repeated complaints about work measurement and time study. They consume a substantial amount of resources. This complaint is partially valid. While the establishment of standards is a somewhat expensive process, by using these standards repeatedly over time in many different areas of work, the benefits greatly outweigh the one-time cost of development. The cost of standard development can be reduced significantly by developing standard data. This fact was realized quickly following the introduction of MTM in 1948.

Standard data can be used to minimize the cost of work measurement in the following situations:

- Work involving constant elements that do not vary from one job to another.
- Elements that are similar in nature but vary due to the level of difficulty of the job involved, the size of the workpiece, the pressure required to be exerted, etc.
- Sometimes, the technical characteristics of materials and processes, can vary such as speed, depth of cut, feed, etc. This can lead to variation in performance time.

When work involves constant elements that do not vary from one job to another, performing numerous studies on the same element can facilitate the compilation of standard data. Similarly, when technical characteristics vary, the standard data for each element can be computed from the physical characteristics of the equipment, such as the length and diameter of the job or the spindle speed and feed rate, which are necessary to perform a machining operation. The greatest impediment to the computation of standard data occurs when one is faced with the task of developing time standards for elements that are similar in nature but vary due to the level of difficulty of the job. The following section describes the method in which standard data are used and applied in real-life situations.

7.2 DEVELOPMENT OF STANDARD DATA

The composition of standard data is in some ways quite similar to the composition of cost. Any engineer conversant in costing analysis will know that there are two elements of cost, namely fixed cost and variable cost. Similarly, there are two elements to standard time data:

1. Constant element
2. Variable element

1. *Constant element*: As the name implies, the value of the constant time element largely remains unchanged from one cycle to the next. Thus, any time element that comprises the action; "start machine" would constitute a constant time.
2. *Variable element*: On the other hand, any time element that deals with the actual machining process itself would constitute variable time, because every machining situation is different. It is based on the product features and machining parameters. For instance, the act of drilling a hole 1 in. in diameter would constitute a variable element. This is because the time required to drill the hole is variable. It changes from one job to the next and depends on several factors such as the depth of the hole, feed rate of the drill, speed of the drill, and the type of material being machined.

As and how they are developed, standard data are indexed and filed. Constant elements are separated from variable elements. Moreover, setup elements are accounted for separately. The standard data for most machining operations are tabulated in the following order:

1. Setup time
   a. Constant time
   b. Variable time
2. Machining time per piece
   a. Constant time
   b. Variable time

Standard data are developed from a large number of time studies that have been performed previously. The process under consideration is studied over an extended period of time. This implies that no process simulations are counted in the development of standard data. Only those processes whose validity has been proven through actual practice are used to develop standard data. Given the dependence on actual time studies, it is important that the start and end points of each element being timed be defined carefully and precisely. This is essential in order to prevent any time overlap between successive elements. One more reason for the precise definition of end points is the fact that time study data are recorded by not just one observer, but by many observers. Thus it provides an objective metric to ascertain the machining time for an operation. If any values in a standard data tabulation are missing, then it is essential to measure them.

Consider the following example. It illustrates the aforementioned principle to compute short elemental times using simultaneous equations.

EXAMPLE: 7.1

Element x: Pick up a small part
Element y: Place the part in a fixture
Element z: Close the cover of the fixture
Element u: Position the fixture
Element v: Advance the spindle

The above elements are times in clusters. It will be appreciated that the individual elemental times are far too short to be accurately timed. We use the simultaneous equations to compute such times as follows:

$$\text{Equation 1: } x + y + z = \text{element } 1 = 0.14 \text{ min} = X$$

$$\text{Equation 2: } y + z + u = \text{element } 3 = 0.134 \text{ min} = Y$$

$$\text{Equation 3: } z + u + v = \text{element } 5 = 0.146 \text{ min} = Z$$

$$\text{Equation 4: } u + v + x = \text{element } 2 = 0.122 \text{ min} = U$$

$$\text{Equation 5: } v + x + y = \text{element } 4 = 0.136 \text{ min} = V$$

It will be observed that there are five variables and five equations. Thus the value of each variable can be safely determined.

Adding all five equations, we obtain the following:

$$X + Y + Z + U + V = 3x + 3y + 3z + 3u + 3v = 0.678 \text{ min}$$

Dividing both sides of the equation by 3, we obtain the following:

$$x + y + z + u + v = 0.226 \text{ min}$$

Thus,

$$X + u + v = 0.226 \text{ min}$$

Therefore, since $X = 0.14$ min, we have

$$u + v = 0.086 \text{ min}$$

But, from Equation 3, $z + u + v = 0.146$ min, the value of z can be determined by simple subtraction:

$$z = 0.146 - 0.086 = 0.06 \text{ min}$$

Likewise, the recurrent application of this principle to other equations yields the following values:

$$x = 0.036 \text{ min}$$

$$y = 0.044 \text{ min}$$

$$u = 0.03 \text{ min}$$

$$v = 0.056 \text{ min}$$

Table 7.1 depicts a form used for summarizing data observed from an individual time study in order to develop standard data in die casting machines.

7.3 DATA IN TABULAR FORM

As we have already seen, machining time is a function of different variables such as the depth of the cut, length of the cut, type of the material being cut, cutting speeds, and feed rates. Sometimes, this necessitates that the analyst has an access to horsepower information as it pertains to the variables being studied. Such knowledge facilitates efficient machining by preventing the machine from being overloaded by suboptimal machining conditions.

Table 7.1: Development Form for Standard Data

Die casting machine

Part no.: __________ Machine no. and type: __________ Operator: __________ Date: __________

Of: __________ No. of parts in tote pan: __________ Method of placing parts: __________

Total weight of flash, parts, gate, and sprue: __________ No. of parts per shot: __________

Liquid metal: __________ Plastic metal: __________ Chill: __________ Skim: __________

Drain: __________ Capacity in lbs. holding pot: __________ Describe greasing:

____

Describe loosening pf part: ____

Describe location: ____

Elements	Time	End points

Consider the example of machining a high-alloy steel forging on a lathe rated at 10 HP. In such a situation, it would not be advisable to initiate a depth of cut equal to 0.375 in. and a feed rate equal to 0.011 in./revolution at a speed of 200 surface feet per minute. Tabular data available from the machine tool manufacturer indicate that from the aforementioned variables, it would be advisable to use a machine tool rated at 10.6 HP. On the other hand, a feed rate of 0.009 in. at a speed of 200 surface feet/min can be accomplished using a machine tool rated at about 8.7 HP. An example of the tabular data used in the above calculation is presented in Table 7.2.

7.4 DATA USING PLOTS

A plot is a graphical representation of tabular data. Data are plotted for two main reasons:

1. Space constraints
2. Provide the analyst with a quick pictorial representation of the trends within the data

Consider Figure 7.1. It represents a plot of forming time (expressed in hours per hundred pieces for a certain part, represented on the y-axis) against a range of sizes (expressed in square inches, represented on the x-axis). Each point on the chart represents an individual time study. It will be appreciated that the points generally seem to follow a linear pattern. It would be possible to fit a straight line to the data; the equation of the line could be expressed by the following formula:

$$\text{Standard time (T)} = 50 + 0.0004\ (\text{area})$$

Table 7.2: High-Alloy Steel Forgings Turning Process: Horsepower Requirements

Surface	3/8 in. Depth Cut (Feeds, in./rev.)						1/2 in. Depth Cut (Feeds, in./rev.)					
Feet	0.009	0.011	0.015	0.018	0.020	0.022	0.009	0.011	0.015	0.018	0.020	0.022
150	6.5	8.0	10.9	13.0	14.5	16.0	8.7	10.6	14.5	17.3	19.3	21.3
175	8.0	9.3	12.7	15.2	16.9	18.6	10.1	12.4	16.9	20.2	22.5	24.8
200	8.7	10.6	14.5	17.4	19.3	21.9	11.6	14.1	19.3	23.1	25.7	28.4
225	9.8	11.9	16.3	19.6	21.7	23.9	13.0	15.9	21.7	26.1	28.9	31.8
250	10.9	13.2	18.1	21.8	24.1	26.6	14.5	17.7	24.1	29.0	32.1	35.4
275	12.0	14.6	19.9	23.9	26.5	29.3	15.9	19.4	26.5	31.8	35.3	39.0
300	13.0	16.0	21.8	26.1	29.0	31.9	17.4	21.2	29.0	34.7	38.6	42.5
400	17.4	21.4	29.1	34.8	38.7	42.5	23.2	28.2	38.7	46.3	51.5	56.7

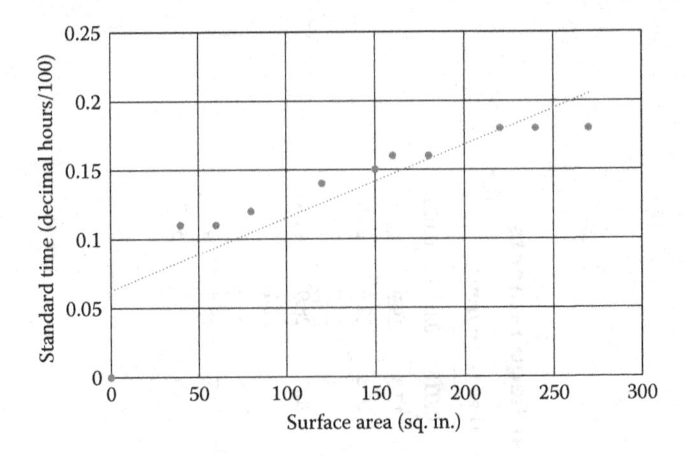

Figure 7.1 Relationship between forming time and surface area.

It will be appreciated that the value of 0.0004 constitutes the slope of the straight line.

As a rule of thumb, at least 10 studies should be available for the construction of a formula to be undertaken. Any number less than 10 will severely hinder the accuracy of the formula. The data used to derive the formula may not be representative either. The amount of data availability is directly proportional to the number of studies undertaken. Thus, it is important to remember that the accuracy of the formula is a direct representation of the data used to derive it.

Once the data are obtained, they are recorded in a spreadsheet in order to analyze and differentiate between constants and variables. The constant time is generally taken as the "y" coordinate on the plot. It is a fixed value and is termed as "the constant." Plotting time against the independent variable facilitates deduction of any algebraic relationships that may be present. Plotted data can correspond to a straight-line relationship, nonlinear increasing trend, nonlinear decreasing trend, or no clearly evident trend. The straight-line relationship is depicted by the following formula:

$$y = bx + c$$

The values of "b" and "c" are derived from least squares regression analysis. Any plot that exhibits a nonlinear trend, power relationships of the form x^2, x^3, x^n, e^x, etc. should be determined. This is true for nonlinear increasing trends. In the case of nonlinear decreasing trends, the values of negative powers of x should be computed. Log relationships, negative exponentials, and asymptotic trends are depicted using the following formula:

$$y = 1 - e^{-x}$$

In this case, the simpler the formula, the better it will be applied and understood. The value of variance needs to be minimized as much as possible. If there is a high degree of variance in the data, then it should be explained. It should be attributed to a specific cause that should be clearly defined in the model formulation.

Once a formula has been determined, it needs to be checked and verified before release. The most effective way to accomplish this is to check the existing time studies. Any significant difference between the time study value and the formula value needs to be investigated (significant being a deviation equal to 5%).

The formula report should include consolidated data, calculations, derivations, and potential applications of the formula. Other factors such as process, operating conditions, and scope of the formula should also be included in the report. The application of plots in order to compute an algebraic formula is demonstrated by the following example:

EXAMPLE: 7.2

Detailed studies on welding operations yielded the following results:

No.	Weld Size	min/in.
1	0.125	0.125
2	0.1875	0.131
3	0.250	0.149
4	0.375	0.241
5	0.5	0.368
6	0.625	0.6
7	0.6875	0.802
8	0.75	0.928
9	0.875	1.139
10	1	1.518

The data plot using Excel yields the result depicted in Figure 7.2.

The first step is to try and fit a straight line to the data. This yields the following equation:

$$y = -0.245 + 1.572x$$

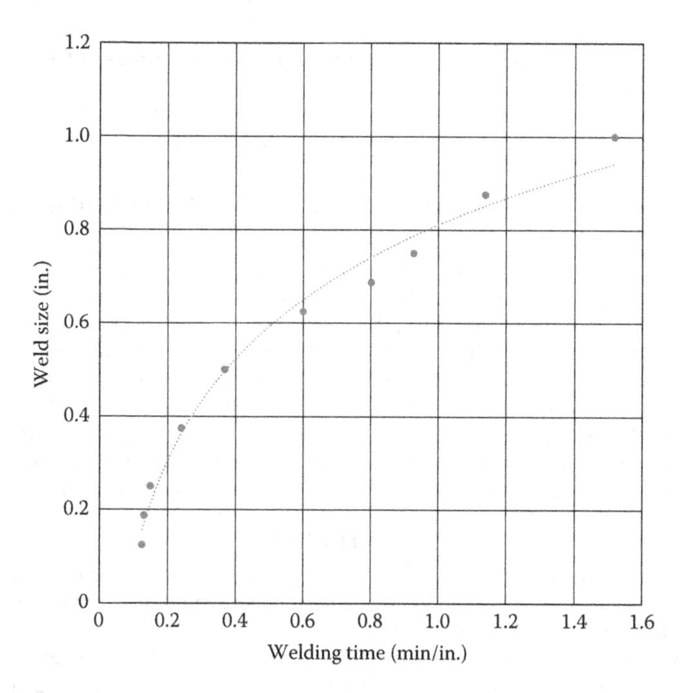

Figure 7.2 Quadratic curve for weld time.

The value of r^2 is 0.928 and the sum of squares (SSE) = 0.1357.

It will be appreciated from Figure 7.2 that the trend is not linear. It is more quadratic in nature. If such a component is added to the equation, the regression equation now reads as follows:

$$y = 0.12 - 0.177x + 1.61x^2$$

The value of r^2 is 0.993 and the sum of squares (SSE) = 0.012.

The fact that not only did the value of r^2 increase, but it is now beginning to approach the value of unity implies that the quadratic model is a far better fit than the previous linear model.

STATISTICAL TEST

Using a general linear test, this improvement can easily be tested statistically as follows:

$$F = \frac{[SSE(R) - SSE(F)]/(df_R - df_F)}{SSE(F)/df_F}$$

SSE(R) = sum of squares for the linear model
SSE(F) = sum of squares for the quadratic model
df_R = degrees of freedom for the linear model
df_F = degrees of freedom for the quadratic model

This comparison yields the following result:

$$F = \frac{(0.1357 - 0.012)/(8 - 7)}{0.012/7}$$

This value is computed as 72.15.

The value of F from the table for one and seven degrees of freedom, respectively, is 5.59.

Since $72.15 \gg 5.59$, it can be safely concluded that the quadratic model is a vast improvement over the linear model.

7.5 APPLYING STANDARD DATA USING ANALYTICAL FORMULAS

Analytical formulas are often found in technical manuals and handbooks. These formulas are applicable to specific machining processes such as drilling, milling, and turning. Cutting times for specific operations and part geometries can be easily computed by ascertaining the appropriate feeds and spindle speeds for different types as well as thicknesses of materials. The following discussion will address each of the three principal machining processes starting with drilling.

The greatest amount of difficulty comes about when timing very short individual elements. The individual values of these elements are determined by timing groups of elements collectively and using the algebraic principle of simultaneous equations to compute the values of individual elements.

7.5.1 Drilling

A drill is generally defined as a fluted end-cutting tool that has one objective: to initiate and then enlarge a hole to a specified dimension in a given material. By definition, the hole is circular in shape.

To compute the total distance traveled by a drill and to finish the entire drilling process, it should be remembered that this corresponds to the length of the hole in addition to the lead of the drill. This is important because it adds to the total amount of distance that the drill needs to travel. The greater the distance, the longer the machining time.

There are two main types of holes: through holes and blind holes. The former type is characterized by a hole geometry that requires the drill to pass entirely through the object. A blind hole, on the other hand, ends within the object. Thus, the length of a blind hole is always less than the thickness of the object being drilled. In this case, the total length of drill travel is equal to the distance between the surfaces of the object being drilled to the point of deepest penetration of the drill.

The following formula is used to compute the lead of a drill:

$$l = \frac{r}{\tan A}$$

l = lead of the drill
r = radius of the drill
A = half the value of the included drill angle
$\tan A$ = tangent value of "A"

For instance, to calculate the lead of a drill 2 in. in diameter, substitute the respective values in the equation. It has to be borne in mind that the included angle for a general-purpose drill is taken as 118°.

Thus, the value of A equates to 118° divided by 2. Therefore, A = 59°.

The value of r is known. This is the radius of the drill. The diameter of the drill is 2 in.; therefore, the radius will be half that value and equates to 1 in.

Substituting these values in the equation yields the following:

$$l = \frac{1}{\tan 59}$$

$$l = \frac{1}{1.6643}$$

$$l = 0.6008 \text{ in.}$$

The aforementioned length is added to the length of the hole in order to ascertain the total distance that the drill must travel.

The total travel distance of the drill is further divided by the feed of the drill (expressed in terms of inches per minute) in order to compute the cutting time in minutes.

Drill feed is expressed in terms of thousandths of an inch per revolution and the drill speed is expressed in terms of feet per minute (fpm). The feed can be converted into inches per minute using the following formula:

$$F_m = \frac{3.82 f s_f}{d}$$

F_m = feed in inches per minute
f = feed in inches per revolution
S_f = surface feet per minute
d = diameter of the drill in inches

To compute the feed in inches per minute of the same 2-in.-diameter drill running at a surface speed of 200 fpm and a feed of 0.015 in./revolution, we would substitute the respective values in the equation to arrive at the following equations:

$$F_m = \frac{3.82(0.015)(200)}{2}$$

$$F_m = 5.73 \text{ in./min}$$

The cutting time for the aforementioned drill to cut through 3 in. of iron casting can be computed using the following equation:

$$T = \frac{L}{F_m}$$

F_m = feed in inches per minute
L = total travel length of the drill
T = cutting time in minutes

Substituting the values in the formula gives us the following result:

$$T = \frac{3 + 0.6008}{5.73}$$

Thus, cutting time equals 0.6284 min.

Please remember that the time stated above only accounts for the amount of time spent in the actual cutting process. It needs to be enhanced using allowances for variations in material thickness as well as for tolerances. Inclusion of these tolerances is important in the calculation of standard time. Personal allowances can also be added onto the existing time in order to account for the human factor in machining.

The following section will deal with formulas and calculations used when computing the machining time in the case of a turning operation using a lathe.

7.5.2 Turning

Turning is an operation that is usually performed on a lathe. It involves a work piece that is mounted on the spindle. The work piece rotates around a central axis and the cutting tool is either stationary or translates along a linear axis in order to remove material from the rotating work piece. A variety of work pieces can be "turned" such as forgings, castings, and bar stock.

In the case of most machining operations, machining speeds and feeds are affected by numerous factors such as the design of the machine tool, the condition of the machine tool, material that is being cut, cutting tool design, coolant used to dissipate heat during the machining operation, and the manner in which the work piece and the cutting tool are constrained.

Most of the formulas that are applicable to drilling are also equally applicable to turning. Thus, the cutting time in minutes can be computed by using the following formula:

$$T = \frac{L}{F_m}$$

F_m = feed in inches per minute
L = total length of cut
T = cutting time in minutes

To compute the value of F_m, use the following formula:

$$F_m = \frac{3.82 f s_f}{d}$$

F_m = feed in inches per minute
f = feed in inches per revolution
S_f = speed in surface feet per minute
d = diameter of the work in inches

Once again, factors such as the diameter of the workpiece, length of cut, feed rate, and spindle speed constitute the main constraints of the problem. Once these constraints are clearly spelled out, it is comparatively easy to calculate the cutting time. The time required to affect a cut, when computed using the above set of formulas can be regarded as the "pure" cutting time. It is important to account for allowances in material properties, thickness as well as personal effects in order to arrive at a more accurate and thus representative cutting time.

The following section will present an overview of how standard data are used in the context of a milling operation.

7.5.3 Milling

Unlike the process of turning, milling involves a machining situation wherein the workpiece is stationary. It does not rotate on the machine spindle. In this case, a multiple-toothed cutter is mounted on the spindle and removes material from the stationary, constrained workpiece. The workpiece, although stationary, is mounted on a bed that translates along the x and the y coordinate axes. Thus, the machining is performed by feeding the workpiece past the multitoothed cutting tools. Milling is used to machine plane and irregular surfaces, cutting threads, slots, and gears.

This is one aspect where milling differs from drilling. In the case of the latter, there is no movement of the workpiece with respect to the drill whatsoever. It is held absolutely stationary.

Once again, as is the case with its "sister" machining operations, milling is a process wherein the speed of the cutter is expressed in terms of surface feet per minute. Since the workpiece is fed into the cutting tool, feed is also referred to as table travel and is expressed in terms of thousandths of an inch per tooth. To calculate the cutter speed in revolutions per minute (rpm), the following formula is used:

$$N_r = \frac{3.82 s_f}{d}$$

N_r = cutter speed in rpm
S_f = cutter speed in fpm
d = outside diameter of the cutter in inches

The feed of the work into the cutter in terms of inches per minute is computed using the following formula:

$F_m = f n_t N_r$

F_m = feed of the workpiece into the cutter in terms of inches per minute

f = feed of the cutter in inches per tooth

n_t = the number of cutter teeth

N_r = cutter speed in rpm

The number of cutting teeth recommended for a specific operation can be computed using the following formula:

$$n_t = \frac{F_m}{F_t N_p}$$

n_t = the number of cutting teeth
F_t = chip thickness

To compute the machining time required for a particular milling operation, it is necessary to first calculate the total distance travelled by the milling cutter. This is very similar to a drilling operation and its inclusion of lead of the drill to compute the total drill travel.

Consider a situation in which the diameter of the cutter is 4 in. The cutter has 24 teeth and the feed per tooth is 0.008 in. The cutting speed is 65 fpm. The length of the work to be milled is equal to 10 in. Assume that the depth of the cut is 0.25 in. The task is to compute the machining time for this problem.

The machining time is computed as follows:

$$T = \frac{L}{F_m}$$

T = cutting time in minutes
F_m = feed in inches per minute
L = total length of the cut

It is clear that the values of all the variables are known except the total length of cut. Since the length of approach and postcut clearance need to be added to the length of work that needs milling, the total distance that must be travelled by the milling cutter is higher than that given in the problem, namely, 10 in.

This additional distance is equal to

$$\sqrt{4 - 3.06} = 0.9695 \text{ in.}$$

Thus, the total distance the cutter needs to travel is given by
L = 10.9695 in.,
Now,

$$F_m = fn_tN_r$$

Therefore,

$$F_m = (0.008)(24)(N_r)$$

But,

$$N_r = \frac{3.82s_f}{d}$$

$$N_r = \frac{3.82(65)}{4}$$

$$N_r = 62.075 \text{ rpm}$$

Thus,

$$F_m = (0.008)(24)(62.075)$$

Therefore, F_m = 11.92 in./min

Substituting this value in the equation for cutting time, we obtain the following result:

$$T = \frac{L}{F_m}$$

$$T = \frac{10.9695}{11.92}$$

Therefore, the cutting time for this particular operation is equal to 0.9203 min.

7.6 USING STANDARD DATA

To use standard data effectively, it is essential to tabulate the constant data elements and file them with the machine or the process. The variable data, on the other hand, can be expressed as an equation or a curve and filed as operation.

EXAMPLE 7.3

Standard data for a certain facility and operation class can be tabulated as well. An example is presented in Table 7.3. The standard time for the entire operation can easily be determined by ascertaining the amount of distance by which the strip needs to be moved. For instance, if the strip needs to be moved 5 in., it corresponds to a time value of 0.103 h/hundred hits.

Sometimes, setup times are combined in order to minimize the time required to summarize a series of elements. Table 7.4 presents the standard setup times for a turret lathe. To compute the setup time, it is essential to know beforehand the tooling to be set up in the hexagonal turret and a square turret. Consider, for instance, a case where the following tooling is required for the square turret:

1. Chamfering tool
2. Turning tool
3. Facing tool

The following tools are required for the hexagonal turret:

1. Two boring tools
2. Reamer
3. Collapsible tap

To calculate the standard time for this setup, the value of the relevant tooling under the square turret column is read. Next, the most time-consuming tooling in the hexagonal turret section, namely tapping, is also read. This corresponds to 69.7 min.

Table 7.3: Blanking and Piercing Operation on a Stock Hand Feed: Standard Data

L (Distance in Inches)	T (Time in Hours per Hundred Hits)
1	0.075
2	0.082
3	0.088
4	0.095
5	0.103
6	0.110
7	0.117
8	0.123
9	0.130
10	0.137

Table 7.4: No. 5 Turret Lathes: Standard Data

		Hexagonal Turret						
No.	Square Turret	Partial	Chamfer	Bore/Turn	Drill	S. Tap/Ream	C. Tap	C. Die
1	Partial	31.5	39.6	44.5	48.0	47.6	50.5	58.5
2	Chamfer	38.2	39.6	46.8	49.5	50.5	53.0	61.2
3	Face/Cutoff	36.0	44.2	48.6	51.3	52.2	55.0	63.0
4	Tn bo grv rad	40.5	49.5	50.5	53.0	54.0	55.8	63.9
5	Face and chf	37.8	45.9	51.3	54.0	54.5	56.6	64.8
6	Fa and cutoff	39.6	48.6	53.0	55.0	56.0	58.5	66.6
7	Fa and tn or tn and cut off	45.0	53.1	55.0	56.7	57.6	60.5	68.4
8	Fa, tn, and chf	47.7	55.7	57.6	59.5	60.5	69.7	78.4
9	Fa, tn, and cutoff	48.6	57.6	57.5	60.0	62.2	71.5	80.1
10	Fa, tn, and grv	49.5	58.0	59.5	61.5	64.0	73.5	81.6
11	Circled basic tooling from above							
12	Each additional tool in square			= 4.2* ________				
13	Each additional tool in hex			= 8.63* ________				
14	Remove and setup three jaws			= 5.9				
15	Setup subassembly or fixture			= 18.7				
16	Setup between centers			= 11.0				
17	Change lead screw			= 6.6			Total setup	______ min

Now, there are three additional tools in the hexagonal turret. Thus, multiply 3 by 8.63 to yield 25.89 min. Adding 69.7 with 25.89 yields 95.59 min.

EXAMPLE 7.4

Consider the cast iron bracket in Figure 7.3. The task is to drill two holes each of diameter 1.0 in. (radius 0.5 in.) at a given location.

Often, in such a case, standard data are not combined, but are rather represented in their elemental form. This enables greater flexibility in developing time standards. For instance, standard data used at a certain manufacturing facility are represented in Table 7.5 inclusive of requisite allowances such as personal delay and fatigue allowance. Given its elemental form, it only follows that the elements can be combined in any order as is deemed appropriate to the part being processed. For instance, it can be used to determine the setup time as well as the processing time per part to drill the aforementioned two holes.

Considering the part geometry from Figure 7.3, it will be appreciated that it is a rather simple part that implies that little to no time is required to study it in depth. Thus, the time required to study the drawing is assumed to be minimal. This implies that the setup element "A" from Table 7.5 will not be included in setup time calculations. Moreover, in accordance with policy, the operation card indicates the drill jig number, drill size, spindle speed, and feed as well as the number of the plug gage.

Similarly, it should be noted that the inspector does not perform part inspection. Instead, the operator is instructed to periodically inspect one in every 10 parts manufactured. The inspection is performed using a go/no-go gage that is provided as the standard equipment. Given the density of cast iron, and the size, volume, and the shape of the part being processed, it is obvious that the part is easy to handle. This implies that the manual inspection process can be easily accomplished.

The operation is performed by adhering to the following steps:

1. Study and investigate the drill jig design.
2. Analyze the job from the perspective of the types of motion and the elements required to perform the job.
3. Compile an element summary as represented in Table 7.6.

Note that the total machining time is equal to 0.278 min. This, however, is exclusive of the actual time it takes to remove material during drilling. The actual drilling time to drill the two holes is noticeably absent from the data presented in Table 7.6 and needs to be included.

The amount of drilling time can be easily obtained by using the formulas mentioned earlier in this chapter. Note that the diameter of each hole is 1 in.

For a 1-in.-diameter drill used to drill cast iron, the following parameters will be used:

- Feed = 0.008 in./rev.
- Surface speed = 100 fpm
- Spindle speed = 764 rpm

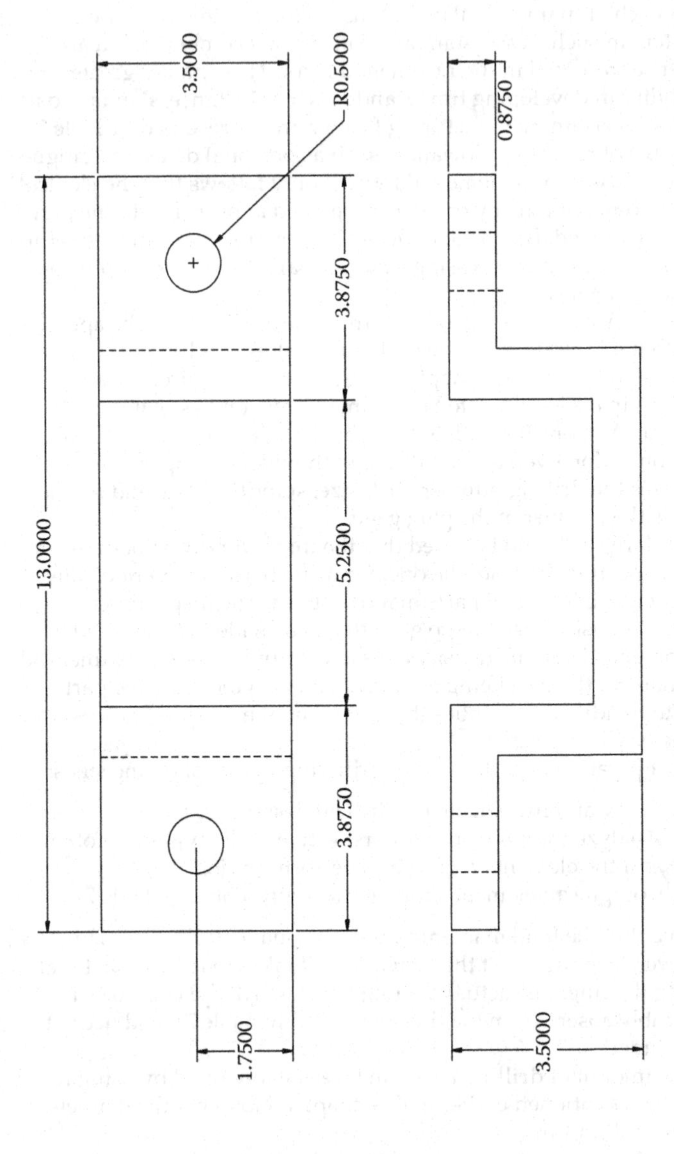

Figure 7.3 Drill two holes, each of diameter 1″ in cast iron bracket.

Table 7.5: Use of Standard Data to Drill Two Holes at a Desired Location in the Bracket

No.	Setup Elements	Time (min)
A	Study the drawing	1.25
B	Obtain materials, tools, and approach workplace	3.75
C	Adjust table height	1.31
D	Start and stop the machine	0.09
E	Inspect the first piece, including normal wait time for inspector	5.25
F	Count production and record it on a voucher	1.50
G	Clean off table and jig	1.75
H	Insert drill in machine spindle	0.16
I	Remove drill from machine spindle	0.14
Element per Piece		
1	Grind drill (prorated)	0.78
2	Insert drill in machine spindle	0.16
3	Insert drill in machine spindle (quick change chuck)	0.05
4	Set spindle	0.42
5	Change spindle speed	0.72
6	Remove tool from machine spindle	0.14
7	Remove tool from machine spindle (quick change chuck)	0.035
8	Pick up a part and place in the drilling jig	
8a	Engage quick acting clamp	0.070
8b	Thumbscrew	0.080
9	Remove part from drilling jig	
9a	Disengage quick acting clamp	0.050
9b	Operate thumbscrew	0.060
10	Position workpiece and advance drill	0.042
11	Advance drill	0.035
12	Clear drill	0.023
13	Clear drill, reposition part, and advance drill using same spindle	0.048
14	Clear drill, reposition part, and advance drill using adjacent spindle	0.090
15	Insert drill bushing	0.046
16	Remove drill bushing	0.035
17	Put part aside	0.022
18	Clean jig (blow air) and put part aside	0.081
19	Inspect part using a plug gage	0.12 for each hole

The spindle speed is computed by plugging in appropriate values into the following formula:

$$\frac{S_f}{\pi d}$$

S_f = surface speed per minute
d = diameter of the drill expressed in inches

Table 7.6: Jig Standard Data Summary for Bracket

Element	Element Description	Time (min)
8	Pk, pt, place jig using quick acting clamp	0.070
9	Remove part and jog using quick acting clamp	0.050
10	Position part and advance drill	0.042
13	Clear drill repeat part and advance drill	0.048
12	Clear drill	0.023
17	Put part aside	0.022
19	Inspect part using plug gage	0.012
1	Grease drill once every 100 parts processed	0.008
2	Insert drill in machine spindle once every 100 pieces	0.002
6	Remove tool from machine spindle	0.001
		0.278 min

An experienced drill press operator will instantly recognize that the spindle speed being equal to 764 rpm is not applicable because most drill presses are calibrated to run at multiples of 100s. If speeds of 700 rpm and 900 rpm are available, we would pick the lower speed to allow a margin of safety in processing the job.

With this piece of information in hand, the drilling time is computed as follows:

$$T = \frac{L}{F_m}$$

The value of "T" corresponds to the time required for drilling one hole. This needs to be multiplied by 2 to compute the machining time for two holes.

The value of "L" corresponds to the length of the hole that needs to be drilled. From the part drawing, it is clear that the depth of the hole is equal to 0.875 in. Clearance for the drill needs to be added to this length in order to compute the value of "L."

The value of the clearance is given as

$$\frac{0.5}{\tan 59}$$

Thus, the value of "L" is equal to

$$L = 0.875 + 0.300 = 1.175 \text{ in.}$$

Next, we compute the value of the feed rate in terms of inches per minute:

$$Fm = (700)(0.008) = 5.6 \text{ in./min}$$

Plugging in the values in appropriate places in the equation for machining time, we obtain the following:

$$T = \frac{1.175}{5.6} = 0.2098 \text{ min}$$

This value is multiplied by 2 in order to compute the total machining time. Thus, the total machining time is equal to 0.4196 min.

The next step is to add an appropriate allowance to the total machining time. Incorporation of a 10% allowance would imply a machining time equal to 0.4616 min. Please note that this is true *only* for machining time, *not* for handling time. That time is added to the machining time. Thus a 10% allowance to the handling time yield the following:

- Total machining time equal to 0.4616 min
- Total handling time equal to 0.278 min as we have seen before
- Total time equal to 0.7396 in order to fully process one casting on one drill press

7.7 SUMMARY

This chapter examined the concept of standard data in detail. Various forms of implementation of standard data were also presented. The method of obtaining and formulating standard data was explained. The use of standard data vastly simplifies managerial and administrative issues in a manufacturing setting, especially those in which unions play a significant role. By alleviating the need to perform actual time studies in order to ascertain standard time, the relationship between management and labor can be improved somewhat by using standard data.

The greatest advantage of using standard data is that a semiskilled worker can determine the standard time for an operation by plugging in values from standard data into formulas. The amount of error inherent in computing standard time using standard data is also quite negligible.

8 Measuring White-Collar Work

8.1 BASIC CONCEPT OF WHITE-COLLAR WORK

The focus of this book is on work measurement. Thus far, the techniques presented in this book deal largely with work measurement as applied to manufacturing activities and an industrial environment at large. This work is not limited to industrial environments alone. A second category of work is becoming increasingly mainstream. Any task that is performed in the service sector of the economy can be referred to as "white-collar" work. This is different from the so-called "blue-collar" work. Most of the tasks accomplished at institutions that are people intensive can be categorized as white-collar work.

Examples of people-intensive institutions include, but are not limited to, banks, utilities, home offices of manufacturing companies, and insurance companies. With the gradual metamorphosis of the U.S. economy into a service economy, white-collar workers are in heavy demand. Examples of white-collar workers include many workers within government agencies, banks, insurance companies, and utilities, etc. A great deal of white-collar work involves a considerable amount of routine work to process. The staff of the aforementioned organizations includes a small percentage of managers and a very large percentage of clerical workers. White-collar workers can pursue two broad career paths: the first path is purely clerical in nature and the other path involves managerial and professional functions. In the modern economy, almost 90% of jobs performed by recent college graduates (both men and women) are increasingly white-collar jobs. This transition to a service economy has largely been enabled by technology.

It will be appreciated that technology has been the major contributor to productivity improvement for many years. Productivity enhancement has been driven by capital investment in better equipment, better tooling, processes, and products. But this is only true insofar as the manufacturing environment is concerned. It has had very little impact on white-collar productivity itself. As a matter of fact, even the impact of computers on white-collar productivity is questionable.

It should be noted that people are the true key to productivity from the perspective of the knowledge worker. Demanding responsibility of them can enhance knowledge worker productivity. Ways and means to try and facilitate the performance of such workers should be sought; time, information, and tools for doing the job should be provided; and any barriers to successful completion of the job need to be removed. The contribution of the white-collar worker to achieving the goals of the organization should be recognized. Thus, considerable investment needs to be done in human resources (HR) in order to enhance white-collar productivity. The relationship between investment in HR and that in capital resources is actually quite similar. Both investments can be evaluated on the basis of return on investment and return on assets. A conceptual framework to measuring white-collar work and productivity is presented in Figure 8.1. In terms of content, Figure 8.1 is self-explanatory. The role of information in enabling white-collar work needs some further elaboration.

Information plays a vital role in an office environment. It provides signals that enable effective functioning of an organization. Information can be formalized in the form of systems and procedures that represent the appropriate manner in which information is to be processed. Such systems and processes can be implemented manually or one can rely on varying degrees of automation to accomplish the task. Either way, the main objective of any system or function

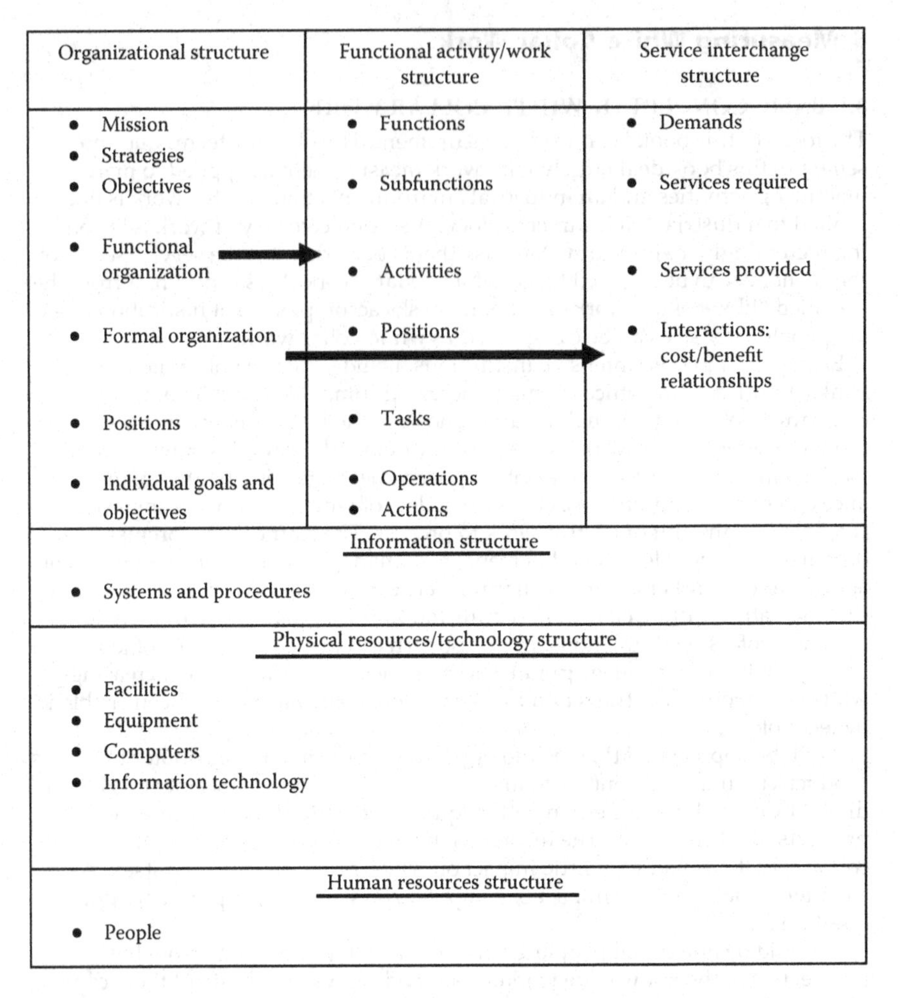

Figure 8.1 Conceptual framework to measure white-collar work and productivity.

is to help workers and facilitate the performance of their job. It is important in an office environment that individual job performance is effectively supported. Such a job should further be relevant in terms of its contribution to objectives of the organization and its mission. Moreover, the system or procedure should be internally effective and efficiently accomplished.

In an increasingly computerized office environment, maintaining the efficiency and internal effectiveness of computerized systems and procedures is the responsibility of experts and personnel in charge of computer services. Yet, these experts do not, as a matter of principle, define the structure of systems and procedures. That is the role of the users is performed with advice from experts with respect to interfacing problems. This enables systems and procedures to be designed with an emphasis on user needs, specifically those related to the style of operation and the opportunity to improve validity, efficiency, and effectiveness.

From the perspective of information structure, participative work simplification constitutes a highly successful approach to enhancing white-collar productivity. In this approach, user groups are trained in systems and procedure analysis. The users then critically analyze selected systems and procedures. The group then develops more efficient, effective, and valid systems and procedures with appropriate advice from experts. Experts then computerize the modified systems and procedures.

In general, the following seven strategies enhance white-collar productivity:

- Develop productivity mindedness
- Use equipment aids
- Increase discretionary content of jobs
- Replace performance appraisals with productivity appraisals
- Provide time management training
- Provide motivation
- Manage productivity by objective

In view of the obvious importance of white-collar work and the need to enhance white-collar productivity, the following sections will deal with how such work can be measured and with the systems that enable measurement and the management of white-collar workers.

8.2 STANDARDS OF MEASUREMENT

To measure anything, it has to be compared with a standard. A standard can be defined as any accepted or established rule, model, or criterion against which comparisons can be made. It is established or setup by an authority as a rule for the measure of quantity, quality, extent, or value. Standards are used in all walks of life. For instance, in winter, how cold "should" it be is an example of comparison against a standard temperature for a given geographical location for a certain time of year. Similarly, how long "should" it take to boil an egg is another example of a standard used in cooking.

The hallmark of a standard is its accuracy. Accuracy has little to do with measurement itself. It is more an economic measure than an engineering measure. There are three types of standards in terms of accuracy: loose, accurate, and tight. In terms of white-collar work measurement, accuracy of a standard can be viewed more in line with fairness of the standard to the employee.

For instance, a loose standard is generous and allows more time than necessary to perform a function. This type of standard is lenient to employees, but unfair to the management.

On the other hand, a tight standard requires an employee to work at a pace that often results in fatigue or is just unattainable. Thus, a tight standard is the antithesis of a loose standard in that such a standard is unfair to the employee, but fair to the management. If employees attained tight standards, management would obtain a greater than 100% return on investment in HR.

Certain kinds of actions increase a manager's chance of succeeding by achieving their ultimate goal whether in a white-collar work environment or in a manufacturing environment. Setting high standards constitutes one of these actions. It has been observed that a manager that sets high standards of performance is often in charge of leading the most highly productive groups of employees. Employees are capable of much more than they realize. One of the greatest problems faced by industrial engineers is to be able to convince a group of employees that a certain task involving a high level of difficulty is actually achievable.

The reverse of the above principle is also true. When standards slip to ever lower levels, where very little has been required and expected of employees over a long period of time, individuals become accustomed to a low output. Furthermore, they are not even aware that they are doing less than a full day's work. If circumstances change and the same group of employees are required to "shape up," the transition is very painful.

In order for standards to be effective, they need to be established scientifically. A scientific standard is equated to a fair day's work pace or 100% performance. A 100% performance is defined as the work pace at which an average, well-trained employee can work without undue fatigue while producing an acceptable quality of work. This corresponds to a "normal" work pace, one that is neither too fast nor too slow. This concept has been presented elsewhere in this book.

8.3 BENCHMARKING WORK

Work can be measured using different benchmarks. Figure 8.1 presents the generally accepted benchmarks of performance when compared with the 100% level of performance. A brief description of each benchmark follows:

1. *50% Performance*: 50% performance is considered to be the average level of performance before any formal controls are applied. This range is usually 40%–60%. The main concept of 50% performance is that employees accomplish about 50% of what could be accomplished once standards are applied to the task at hand. It does not imply doing a half-day's work.
2. *70% Performance*: 70% performance is regarded as the minimum tolerable level of performance. In other words, trained employees should be able to perform at this pace posttraining. If not, it would be inadvisable to let them continue in that capacity since this implies nonoptimum use of resources invested in that employee.
3. *100% Performance*: 100% performance corresponds to a fair day's work level. This is considered to be the goal of all employees. It is also the goal that all supervisors and managers seek to achieve. It is to be noted that in spite of undergoing training, not all employees have the aptitude to actually achieve 100% performance in their daily job function. Maintaining a high level of performance everyday requires a high level of perseverance, concentration, and dedication. Not all computer programmers are able to "code" at the 100% level every single day. The same can be said for typists, etc. Given the practical impossibility of attaining and maintaining this level of performance, it is advisable to use the 85% performance level as the start of an acceptable range for performance measurement.
4. *120% Performance*: 120% performance level is considered to be the incentive pace. This is because most incentive programs are based on the concept of a 20% bonus. It also means that employees can exceed the 100% level by 20% on average in the presence of the provision of an incentive or award. Employees working at this level are often rewarded in some form either in the shape of direct reward or compensation.
5. *135% Performance*: 135% performance level corresponds to the expert level of performance. It can only be achieved and maintained by a handful of expert

employees or by extremely skilled workers. It requires an unusual mental acumen and/or physical dexterity to be able to perform work at this pace and maintain it (Table 8.1).

8.4 AN APPROACH TO MEASURE WHITE-COLLAR WORK

Arguably, the mental caliber of white-collar employees is usually higher than that of their blue-collar counterparts. The nature of the clerical type of position is also quite different than that of a technical position. Any effort to measure white-collar work should take these two attributes into consideration. A clerical work measurement program should be different from a technical work measurement program in terms of the program name, record keeping requirements, design of documentation, reporting techniques, approach to setting standards and communication, and employee participation.

Similar to a technical work measurement program, a clerical work measurement program has several essential components:

1. *Specific communication*: The program must be introduced and implemented systematically with a carefully thought out plan. In case of a clerical work measurement program, a few short meetings are generally sufficient to achieve the objective. The following content should be addressed in adequate detail during the meetings:

   a. *Feasibility study*: The analyst meets with the HR manager to explain the objectives of the study such as ascertaining the nature of work, patterns of work completion, output activities, and processing problems. Results of the study and the accompanying recommendations are then reported to the management.

   b. *Meeting with employees*: The results of the feasibility study and objectives of the program are discussed in detail and well in advance of program implementation.

   c. *Conducting the study*: The objectives of the study and approach are explained to all employees being evaluated with a particular emphasis on employee participation.

   d. *Task list review*: Each employee is interviewed and a task list is developed with his or her approval.

   e. *Procedures*: Each interview with an employee enables the analyst to document procedures and have them reviewed for approval. Changes, if any, are incorporated at this stage. Any such changes should not be incorporated without a thorough discussion.

Table 8.1: Performance Benchmarks

150%	Super skilled level work
135%	Expert pace work
120%	Incentive pace work
100%	Fair-day's work pace
85%	Acceptable level
70%	Minimum tolerable level
50%	Normal level before controls

f. *Frequency study*: Each step performed by the employee must be assigned a relative value expressed as a percentage of the key activity. These percentages are ascertained with the help of employees through a frequency study performed over a sufficient period of time. By definition, the period of time should be long enough in order to be statistically valid and be representative of the cycle time for the activity under consideration.

g. *Presenting the standards*: This step involves an explanation on the part of the analyst in sufficient detail as to how the standard was developed.

h. *Record keeping system*: The mechanics of the record keeping system, means of capturing data, result display mechanism, and the usability of the resulting data is explained in this step.

2. *Employing analysts having specific expertise*: Employment of analysts specifically trained to perform this type of study that tends to increase the confidence level of the employees. It is felt that the complexities of the job are well understood and fairly evaluated. Owing to familiarity with the study on the analyst's part, the length of such a study is actually reduced. A proper perspective can be maintained and the study is widely accepted.
3. *Emphasizing participation*: Similar to its blue-collar counterpart, white-collar work measurement emphasizes employee participation. Employees are interviewed in detail and their help is enlisted in structuring the job procedures. Any deviation from the normal method of performing a job is discussed and resolved. It is imperative that everyone supports the new procedure.
4. *Documentation*: Most jobs are considered to be complete procedures by employees rather than fragmented tasks. Any documentation pertaining to the job should be recorded in sequence taking care to include all steps routinely performed and any exceptions that may deviate from the routine.
5. *Record keeping*: Any system that aids in the keeping of records needs to be straightforward and uncomplicated. Above everything else, record keeping should not hinder the employees' ability to effectively perform their job. Volume counts should be limited to about one item. It should not be more than three in any case. Meetings and interruptions should be accounted for and there should be no need to record them as off standard time.
6. *Reporting period*: The reporting period for most programs needs to be based on the logical cycle of completing the entire procedure from beginning to end. This entails a reporting period at least every 2 weeks or monthly once.
7. *Using the results*: Goals should be appropriate to the type of job being analyzed, given the large variety of white-collar jobs. Results should always be evaluated in terms of ranges. Percentage efficiencies should be computed; however, they are more meaningful when considered in terms of a range rather than individual values. Thus, 85%–95% is more significant than just say 87%.

There are three main techniques of measuring white-collar work, namely informal techniques, semiformal techniques, and formal techniques. Each type of technique will be dealt with in some detail in the following paragraphs. Table 8.2 compares different work measurement techniques based on the level of "formality" involved in measuring work.

Informal techniques: Informal techniques commonly use "estimates" to measure work. An estimate can be defined as an approximate judgment of amounts. Most

Table 8.2: Comparison of Work Measurement Techniques

Technique	Accuracy	Consistency	Control	Savings
Formal	High Can be validated by other formal techniques	High Properly trained analysts working independently will arrive at same standards within 5%	High Due to factual basis of information	High (20%–45%)
Semiformal	Moderate Will not hold up under validation	Moderate Savings will vary about 10%	Moderate Due to lack of attention to methods and work pace	Moderate (10%–12%)
Informal	Low	Low No consistency Could vary 50% or more	Low Due to reliance on opinions rather than facts	Low (1%–5%)

estimates are largely personal in nature and are based on a person's skill to estimate. The amount of subjectivity inherent in estimates and "estimating" brings up the issue of accuracy as well as consistency. The process of estimating costs takes almost nothing to accomplish and requires no special training other than a degree of familiarity with the jobs being estimated. Most estimates, therefore, cannot be supported. Estimates are loosely based on the method prevalent at the time the estimate is made. The informal nature of the estimate implies that the method is seldom documented, thus rendering any changes in it to go unnoticed. Moreover, the subjectivity element inherent in informal techniques makes it hard to defend a particular estimate.

Semiformal techniques: Informal techniques of work measurement represent an attempt to compute the average amount of time it takes to process work. Semiformal techniques, on the other hand, seek to separate the time devoted to a task from nonproductive time such as personal time, idle time, and leaves of absence. This is also referred to as work sampling, a concept that was presented in Chapter 5.

Formal techniques: Formal techniques of work measurement adhere to formalized rules and algorithms for measuring work. They can be distinctly categorized as time study and predetermined time data, each of which has been dealt with in previous chapters. They are regarded as engineering work standards. Owing to their formal and rigorous nature, such standards are heavily relied upon to measure individual performance. Wage incentive plans are also based on such standards.

It will be appreciated from Table 8.2 that formal techniques of work measurement are inherently more controllable, exhibit a greater degree of consistency, and can be validated by other formal techniques that results in a substantially high savings rate ranging from 25% to 45%. It is no wonder that such techniques are relied upon to measure work in an office setting.

Most white-collar work is measured using some derivative of the MTM system described in detail in Chapter 6. The remainder of this chapter will present such methods in detail. The concepts of office standard data, office standard data systems, multiple linear regression (MLR) analysis, and management by objectives will be discussed at some length.

8.5 OFFICE STANDARD DATA

The concept of standard data and MTM has already been discussed in Chapter 6. This concept can be further extended to include white-collar work or office work. One factor that should be taken into consideration is the amount of time required to measure office functions. The amount of time required to develop standards is directly proportional to the size and the number of work elements. Therefore, it is important to reduce the number of work elements and to increase the length of time of such elements. This step is referred to as "developing standard data." Almost all standard data are established in a stepwise manner. The central concept is that progressively larger time elements can be derived for rapider and less expensive direct rate setting. This can only be accomplished by accepting some loss of accuracy as well as reduction in methods improvement capability.

Think about the case of a simple office function—stapling papers together. This is depicted in Table 8.3. It will be observed that the task requires the analysis and recording of seven distinct MTM elements. The duration of the task is about 41 TMU. The seven different MTM elements used in the establishment of the standard constitute building blocks that are almost always repeated in some order every time the main task is performed. Once the standard is fully developed it need not be reanalyzed to compute standard time for each iteration of the task under consideration. In other words, it can be predicted with a high degree of accuracy that stapling a bunch of papers will always consume about 41 TMU.

It needs to be borne in mind that all standard data elements must be developed in such a way as to be uniformly applicable every time the operation in question is performed. In this specific case a high degree of predictability is necessary in the analysis of the following elements of operation:

1. Manner in which the sheets of paper will be picked up each time prior to stapling
2. The location and manner in which the material will be set down prior to stapling
3. The location and manner in which the material will be set down after stapling
4. The manner in which the material will be picked up and set aside every time one staple is affixed to the batch

Table 8.3: Application of Standard Data to Estimate Time to Staple Sheets

Description Left Hand (LH)	Symbol	TMU	Symbol	Description Right Hand (RH)
		15.2	M 12C	To stapler
		14.7	P1SSd	Align
Hit	R5Am	5.3		
	G5	0.0		
Depress	mMIA	1.9		
	RL2	0.0		
Hand away	R2E	3.8		
		40.9		

If a high degree of predictability can be incorporated into the above sequence, then they can be included in a larger motion pattern. If the aforementioned sequence and individual actions lacks a degree of predictability, a data set developed for stapling items together should address the following parameters:

1. Start with the items to be stapled in hand
2. Move the stapler, position the items to the jaws of the stapler, inset the items, and staple the material
3. End with the stapled items in hand

It should be noted that all standard data could be classified as either horizontal standard data or vertical standard data. Vertical standard data are based on actual work task elements. It is therefore restricted to only one kind of work. Horizontal standard data are based on motion sequences that are common to many types and classes of work. Some basic office standard data systems will be presented in Section 8.6.

8.6 OFFICE STANDARD DATA SYSTEMS

Office standard data systems draw upon the concept of MTM, which is a predetermined time system that is commonly used in a manufacturing setting. For a detailed description of the structure and usage of MTM, refer to Chapter 6. This section of the chapter will deal with MTM application in an office setting.

There are three levels of horizontal standard MTM data represented as follows:

- *Level 1 data*: Level 1 comprises the basic data such as the MTM. It includes time values for basic motions that are performed by the body. Examples of such motions include reach, bend, move, sit, etc.
- *Level 2 data*: Level 2 data is developed using level 1 data. Level 2 data can be regarded as a derivative of level 1 data. Thus, actions such as picking up a sheet of paper, pressing a key on a keyboard, and writing a letter are typical office functions that can be characterized as level 2 data.
- *Level 3 data*: Level 3 data is developed from either level 1 data or from level 3 data. Examples of level 3 data include actions such as opening mail, typing an entire document, and addressing an envelope. It will be appreciated that level 3 data draws upon level 1 as well as level 2 data. It can be regarded as a more through expression of level 2 data.

Numerous standard data systems have been developed based on the MTM format. They are mostly level 2 data systems. Examples of some of these systems are presented below:

- *CSD*: CSD is an acronym for clerical data systems. It was developed by Bruce Payne Associates in 1960.
- *MCD*: MCD is an acronym for master clerical data. It was developed in 1958 by Serge A. Birn Company.
- *UOC*: UOC stands for universal office controls. It was developed in the 1950s by H.B. Maynard and Company.
- *MODAPTS*: The modular arrangement of predetermined time standards was developed by Chris Heyde at Unilever in Australia. It is widely available and has been discussed at length in Chapter 6.

- *MTV*: Motion time values was developed in the 1960s by Booze, Allen, and Hamilton.
- *MTM-C*: MTM-C is a special adaptation of the basic MTM system to suit clerical applications. Thus, it stands for MTM—clerical. It was developed in 1978 by a consortium of MTM association members and is available from the MTM Association of Standards and Research.
- *AOC*: AOC is an acronym for advanced office controls. This level 3 system was developed in 1973. Despite being a level 3 system, it contains a sufficient amount of level 1 and level 2 data in order to accurately perform all types of functions, whether they be repetitive or nonrepetitive in nature. It was developed by Robert E. Nolan Company and will be described in detail in Section 8.6.1.
- *Mulligan system*: The Mulligan system utilizes motion picture analysis. Thus it is comprised mainly of level 1 and level 2 data. It was developed by Paul B. Mulligan Company.

In view of the large number of systems available to measure white-collar work, the question arises as to which system is better. Which system would offer the best advantage over others? Several studies have been conducted to answer these questions. Frequently, these studies show that one system may offer certain advantages such as higher speed of application, broader range of data, more updated office-equipment data, higher degree of accuracy on short cycle operations, or less amount of time required for training. However, it is important to maintain discretion when evaluating the results of such studies. This is because each advantage offered by a system is often offset by several disadvantages, thus making the final decision to be much more difficult. It needs to be pointed out that the final decision needs to be made based on an objective set of constraints.

8.6.1 Advanced Office Controls

As referred to in Section 8.6, AOC is a proprietary system that was developed by Robert E. Nolan Company in 1973. It is an analysis and measurement technique used to control office costs. In its most basic form, it is a library of engineering standard time values that encompass almost every aspect of any type of work that is usually performed in an office setting and its findings are summarized on one card. It contains three levels of data under one coding system. As a system, AOC is economical in its application and offers a high degree of accuracy as well as consistency. It is also quite easy to learn and simple to use.

The primary aim of AOC is to develop a set of data to cover virtually every function that is performed in a productive situation in any office. It can thus be characterized as a case of horizontal standard data. It will be appreciated that a block of data, once developed, is not quite economical to use until it has been coded. The process of coding enables the data to be cataloged and retrieved from memory on command. Some data systems contain a large number of elements thereby hastening their obsolescence. This is because trying to retrieve the previous data from that system becomes very time consuming and thus counterproductive. It would be easier to just develop a new standard from scratch. This is in line with the principle, "It is not possible to economically use an element if that element cannot be found economically."

There are two main types of coding systems: alphabetical and numerical. The alpha-mnemonic coding system is by far the most creative. The term "alpha-mnemonic coding" implies that the system is alphabetical in nature and recalls the memory (mnemonic), thus simplifying the retrieval of coded data. It is based on the following three rules that are easy to follow:

1. No code symbol should comprise more than three letters of the alphabet
2. Each code letter must be the first letter in a word that describes all or part of an element
3. No alpha letter can be used more than once in the first field or more than once in a subcategory in the second or third field

AOC is an extension of the MTM system of establishing time standards. MTM is widely used in a manufacturing environment; however, the scope of its application is quite limited in an office setting. This can be attributed to the comparative dearth of repetitive tasks in an office. To surmount this difficulty, AOC visualizes work in an office as a cycle of action. This is quite different from considering work as individual motions. A cycle of action consists of three main parts:

1. Start the action
2. Perform change
3. Stop the action

Consider a common office function such as opening an envelope. Let us analyze this relatively simple function using AOC.

The action starts when the opener is moved to the envelope. The actual opening of the envelope effects change and the action is stopped when the envelope is completely open. This sequence of actions is presented in Table 8.4. Applying AOC to this process would result in using the code "PEOS" in order to record and standardize the task. This code can be translated as follows:

P (paper handling) of E (envelope) to O (open) a S (sealed envelope)

The total amount of time required affecting this process amounts to 49 TMU.

Table 8.4: Application of Standard Data to a Common Office Function: Opening an Envelope (Levels 2 and 3)

Description	MTM Convention	Time Units
Level 2 Data		
Move opener to envelope	M8C	11.8
Opener to corner of envelope	P2SD	21.8
Insert opener in envelope	M2B	4.6
Slit open envelope	M8B	10.6
		48.8 TMU
Level 3 Data		
Get envelope and opener (later aside)	GMG	36
Open envelope	PEOS	49
Remove contents	GLG	49
Unfold papers	PU(2)	64
		198 TMU

Table 8.4 is split into two categories. The top half analyzes the action of opening a sealed envelope using MTM data. Thus, the top half of the table constitutes the AOC code; PEOS and has a value of 40 TMU. This is what is referred to as level 2 data.

Level 3 data is derived from level 2 data. This is presented in the bottom half of Table 8.4. The bottom half of Table 8.4 presents a complete mail opening operation. Once again drawing upon the cycle of action principle, we have the following steps:

1. Start the action with the picking up of the envelope and opener
2. Affect change with the opening of the envelope, removing the contents, unfolding the contents, and setting the materials on a desk
3. Stop the action by setting the envelope aside

As before (level 2 data involving the code PEOS), a single AOC code, namely GRSF records and standardizes this block of data. GRSF is translated as follows:

G (gather), R (receive), S (sealed envelope), F (folded contents)

The total amount of time required to affect this process is 185 TMU.

It should be pointed out once again that all AOC elements are cycles of action. Therefore, only 226 are required to represent and account for all office work. The average time value is a little more than 15 s, thus relieving the analyst of detail work. It is noteworthy that such facility is achieved without compromising the level of accuracy required for sound management control. It should also be noted that the average time value for the AOC system (a little more than 15 s) is actually more than 55 times greater than that for MTM.

An alphabetical coding system can be used with a great degree of facility. This is the reason AOC relies on such a system instead of a numeric version. Furthermore, an alphabetical coding system can be structured to have a direct correlation with the words describing the basic elements. Thus, words can be memorized easily after only a small amount of exposure due to the high level of logic and precision incorporated into the system.

It has been demonstrated from Table 8.4 that AOC is a third level data system. Thus, it retains the characteristics of level 1 as well as level 2 data within one system. This enables the analyst to choose the appropriate level of data in order to economically establish standards with a particular degree of accuracy. Additionally, AOC also contains a machine data supplement that is updated on all the latest office equipment and machines.

A major advantage of using AOC is the larger size of AOC elemental time values. This means that fewer elements are required to establish standards. Each element provides precise documentation for each task, thus raising the degree of versatility of the system. It will be appreciated that the AOC code and structure are quite simple and straightforward. Supervisors and employees can easily comprehend the basis on which standards were developed. It should also be pointed out that AOC documentation is not limited just to establishing time standards. It is an excellent tool that can be extended to methods analysis as well. It will be recalled from previous chapters that method analysis is widely used to simplify an existing system or to model a proposed system. Since time is the basis for most work measurement, operation cost is also expressed in terms of the amount of time taken to finish a task, AOC can be easily used to perform costing and equipment analysis as well.

8.6.2 Advanced Standard Data Systems

Auto AOC is the name of an automated system of standard data that has been developed for office use. It is a complete package of computer hardware and software. The following are the main objectives of auto AOC:

1. *Task outline*: This objective involves written procedures that document the task and are specific enough to be used in systems analysis and training.
2. *Task analysis*: This objective involves documenting the method by which the standard was developed for each step with the procedure.
3. *Task summary*: This objective places each major procedure in proper relationship with others and computes the standard.

As always, trained analysts and engineers gather facts about the work by interviewing employees. The task content is studied in detail and improvements are made to simplify and streamline the method. The approved method of performing the task is documented in detail. This is used as a basis for computing the amount of time required to perform the work. This task is facilitated by AOC. The amount of time required to perform the task is defined as the standard for that task. Documentation of the procedure for the user departments is recorded and it is retained in order to perform methods analysis, training of new employees, referral, future audit, and update.

Typically, analysts using AOC spend about half the time used to develop standards in writing procedures and setting standards. Almost 80% of the analyst's effort is spent on the standard setting process in companies using first- or second-generation standard setting systems. To summarize, the following advantages are offered by auto AOC:

1. Auto AOC results in reduced analyst effort. This is as much as 20%–55% less than the manual standard setting systems.
2. Auto AOC helps in setting mathematically correct standards.
3. Auto AOC enables storage of standards on diskettes and other electronic storage and retrieval systems, thus facilitating easy retrieval and maintenance.
4. Auto AOC eliminates the need for manual office support activities such as secretarial services and typing procedures that are expensive.
5. It improves analyst morale and reduces turnover due to elimination of routine work.
6. It does not need any data processing assistance for installation and use.

8.7 MULTIPLE LINEAR REGRESSION

MLR analysis is used to help answer questions concerning staffing issues such as indirect labor. Some examples include the following:

- How many indirect workers do I need to run my organization/department?
- How many scientists and related personnel such as engineers, physicists, and chemists do I need in order to successfully perform research?
- How many personnel do I need in product design and service planning?
- How many personnel are required in process development and service planning?

- How many personnel are required to staff other departments such as quality control, and scheduling?
- How many clerical personnel are required to support all functions?

There has been an increasing shift of enterprises globally toward at least a partial degree of automation. In light of this situation, the importance of such questions is obvious. A greater degree of automation implies less reliance on direct labor and an increased reliance on indirect labor. Thus, the ratio of indirect to direct labor is on an upswing. Most direct labor can be classified as "blue collar" whereas indirect labor, by definition, is a support function and can be classified as "white collar."

MLR analysis is an extension of the mathematical process of averaging. A certain degree of variability exists in almost every engineering function. For example, although the standard time to perform a task is constant, the actual amount of time taken to perform that task typically varies. A sufficiently large sample of machining times (time required to type a document, in the clerical context) will show a distribution of data, not a constant value. The arithmetic mean represents the central tendency of a dataset wherein variability occurs due to random chance. It can be counted upon as a good predictor of individual events only when the degree of variability inherent in the data is small. Thus, the arithmetic mean is always used in conjunction with standard deviation. Yet, an increase in variability reduces its accuracy. Moreover, variability is a result of cause and effect and not chance that renders such a measure obsolete. Conditional averages are used to predict the occurrence of events in cases like these as long as the underlying causes or condition of change are specified.

Regression analysis is a process used to estimate the rate of change of expected value relative to changing conditions. For instance, the amount of indirect labor required to accomplish a task can be represented by the notation: Y_i. The subscript "i" denotes the task under consideration. Consider an assignment in which a facilities layout is being developed. Such layouts are a direct function of the floor area under consideration. The final outcome will change based on changes in the floor area, which constitutes the variable condition or predictor variable.

The conditional average $\hat{Y}_i$ is an estimate of the expected amount of labor required in light of the ith set of conditions, the area for the ith layout. This relationship is represented as a straight line. The expression for a simple, one-predictor model is represented as

$$\hat{Y}_i = b_0 + b_1 X_{i1}$$

b_0: X intercept representing the average effect of all potential predictors that are not part of the model

b_1: Regression slope expressed in terms of units of Y per unit of X1

X_1: Predictor or independent variable

The intercept can also be regarded as the amount of time required to perform actions that only happen once per task. Thus, setup time can be regarded as the X intercept.

The regression analysis picks values for the X intercept and slope that corresponds to minimal residual variance. The notation represents the residual variance: $S^2_{Y \cdot X}$. The value is computed by computing the sum of squared deviations of individual points along the regression line. Regression analysis can be likened to a type of least square estimator, similar to the basic concept of arithmetic mean. The sample residual variance is expressed by the following formula:

$$S_{Y \cdot X}^2 = \sum_{i=1}^{n} \frac{d_i^2}{n-2}$$

The residual standard error is the square root of the variance expressed above. It is the standard deviation of individual values around the model. The sample residual variance is a measure of model accuracy in the case of linear regression analysis in the same way as standard deviation is a measure of accuracy of an arithmetic mean. One way to check the accuracy of the model is to check the size of $S_{Y \cdot X}$. Answers are sought to the following questions:

- Is the standard deviation small enough as desired for the intended use of this model?
- Is the reduction from S_Y to $S_{Y \cdot X}$ adequate?

Sometimes, the standard deviation is too large, suggesting that the simple model is inadequate or that the model in its simple form (with only one predictor) cannot fully describe the situation and is incapable of making a prediction. In such a case, we resort to additional predictor variables. The process is continued till the variance is minimized. The incorporation of additional variables transforms the model into an MLR model.

The general form of the multiple regression models is presented in the following equation:

$$\hat{Y}i = b_0 + \sum_{j=1}^{p} b_j X_{ij}$$

If $p > 2$, the equation is considered to be an MLR. Note that the unknown constraints (b_j) appear in a linear form (power = 1). It is therefore referred to as a linear equation.

Before developing an MLR model, it is important to learn about the underlying system in detail. Different sets of variables need to be studied in detail in order to ascertain their potential role on the mathematical model. Similarly, their role in reducing residual variability needs to be studied. The system comprises a large set of true predictors. Any variation in predictors creates variation in response. Real-world applications make it almost impossible to sort out each and every variable and account for all the variation. Instead, the system focuses on selecting the most important variables to be included in the model. A large number of smaller contributors, on the other hand, are left behind as residual error. This process is represented graphically in Figure 8.2.

Contributions to variation in the response variable accumulate from very small at the right to largest contributor at the left. Considering the largest contributor at X1 reduces the variance by 0.33%, the next by 0.25%, the following by about 16%, and so on. Where do we stop? The cutoff point depends on the end goals, availability, and cost at which data can be obtained and on the interrelationship between the predictors. It is to be borne in mind that this is an idealized representation. In this situation, the top four largest contributing predictors were chosen and included in the model, while the remaining were discarded and counted as residual error.

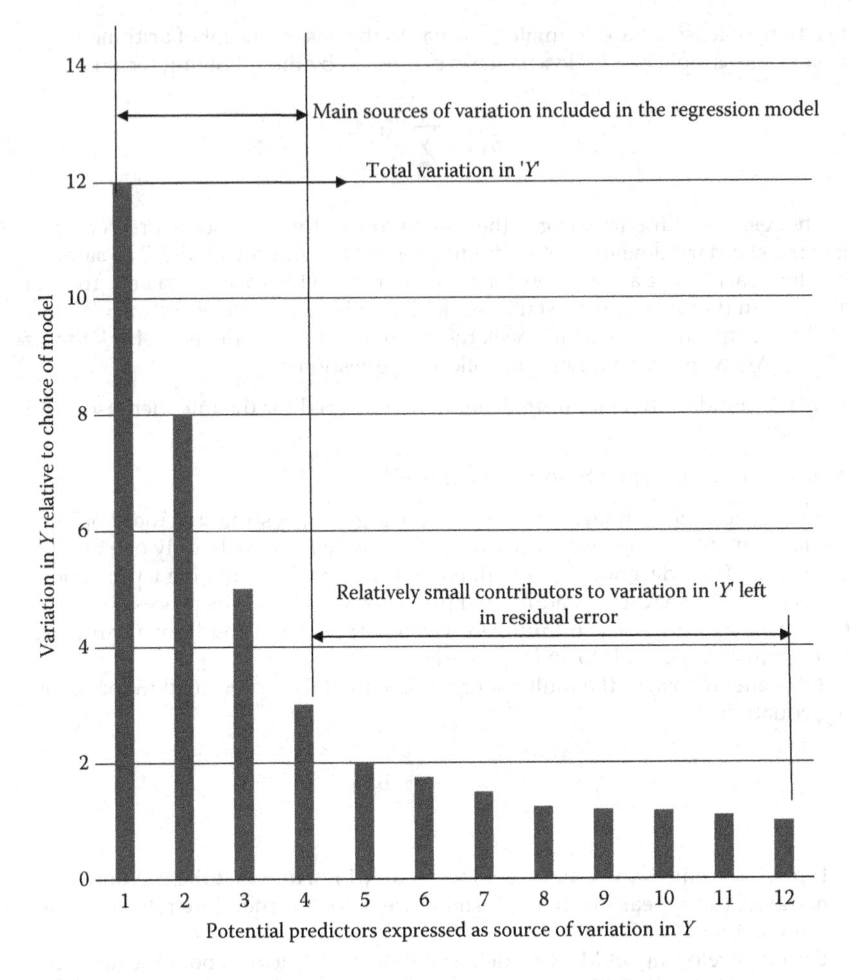

Figure 8.2 Sources of variation ranked according to relative importance.

There is a significant difference in modeling direct labor jobs and indirect labor jobs. In the former, it is essential to use several predictors in order to accurately model the job. The number ranges from 3–4 to 10–12. In the case of the latter, given the greater degree of complexity and variability of indirect labor jobs, approximately at least 5–6 predictor variables are used. The general principle to remember is that the model needs to be suitable, not simple.

In terms of goals of indirect job modeling, there are four main categories of goals enumerated as follows:

1. Exploration: To learn about the system
2. Specification: To pick a model structure
3. Estimation: To ascertain the predictor variables with adequate precision
4. A. Prediction: To use the model in order to anticipate future events

   B. Control: To use the model in order to alter system performance

It should be noted that in a large number of scenarios, not all projects mature to level 4. Each of the aforementioned goals can serve as an end goal by itself. In the case of indirect labor analysis, most projects do seem to evolve to level 3. This is because the main objective of the study is to "estimate" indirect labor cost. Mere exploration and specification constitute an insufficient goal in such cases. Indirect labor modeling has goals that fall into the last three categories as presented here:

1. Estimation: Cost accounting and economic evaluation/planning
2. Prediction: Planning for purposes of recruitment, training, providing facilities, etc., allocating, bidding, and scheduling
3. Control: Changes in trends, detection in trends, and outlier or deviant performance

Remember that many projects have multiple goals in overlapping categories. It is desired to help and clarify the process through the provision of a basic structure that can be scaled as required.

8.7.1 Ambiguous Nature of White-Collar Jobs

Most white-collar jobs can be considered to be situated on a multidimensional space that is ambiguous. The ambiguity exists with respect to job description or function specification. This is almost the exact opposite of structure. In light of the amorphous nature of most white-collar jobs, it is often difficult to determine staffing requirements for such positions. Most white-collar personnel are known to be self-directed that have a talent for knowing and doing what is required without being told what is needed.

Table 8.5 represents the nature of the white-collar job space and the goals and assessment processes related to it. At the top of the table is the job type continuum as it slides across from direct jobs at the extreme left to indirect jobs at the extreme right. Several closely related and overlapping gradations of job characteristics are depicted below with respect to the direct–indirect spectrum. It will be appreciated that a direct job is one that is very structured, relatively uncomplicated, comprising predetermined steps, exhibits a low degree of variability in terms of content and time per cycle. Examples of such jobs include small parts assembly, press feeding, packaging, keypunch operation, assembly line stations, etc.

At the other end of the spectrum are jobs that are complex and unstructured. They require attributes such as self-starting, creative behavior, and problem solving skills. Each task or assignment may be unique in terms of requirements. Simultaneous or overlapping attention to several projects may be required. Each of these projects could last several months or even years. It is obvious that the quality and appropriateness of such an activity is difficult to judge. Examples of these activities include medical doctors, engineers, lawyers, architects, project coordinators, teachers, and statistical analysts. A large number of jobs fill the continuum in between the extremes. Examples of such jobs include sale personnel, nurses, some warehousing and marketing jobs, computer programmers, etc.

Direct assessments of individual tasks are possible and economically practical in the case of a direct job. This can be achieved through the use of time study, predetermined time standards, and other techniques that have been dealt at length in this book.

As the level of job complexity, variability, and cycle duration rises, it becomes less economical and physically less possible to directly assess job performance. In this situation, one needs to resort to indirect means of assessing individual

Table 8.5: Regression Goals Related to the Job Continuum

Job Type	Direct	(Continuum)	Indirect
Ambiguous job	Highly structured Generally unstructured		
Complex job	Low High		
Creative job	Predetermined steps Self-determined steps		
Variable job	Low–high when repeated		
Repetitive job	Normal Tend not to repeat		
Task duration of job	Short Long		
Input–output association	Close, immediately accessible results Remote, results difficult to access		
Assessment process	Direct assessment of individual tasks	Indirect assessment of individual tasks	Indirect assessment of aggregate tasks, functions
Regression role	Generalize results from direct assessment	Assess relationships of labor consumption to measures of production, task descriptions	Assess relationships of staff group sizes to major task indices
Modeling Goals			
Estimation	Cost accounting, economic evaluation and planning, budgeting, and wage determination	Cost accounting, economic evaluation, planning, and budgeting	Economic evaluation and planning, budgeting, assessment of impact of societal changes
Prediction	1. Planning 2. Scheduling 3. Sequencing 4. Expediting	1. Planning 2. Scheduling 3. Sequencing 4. Expediting 5. Bidding	1. Planning 2. Allocation
Control	Detection of the following: 1. Trends 2. Changes 3. Deviant performance above and below expectation	Detection of the following: 1. Trends 2. Changes 3. Deviant performance above and below expectation	Detection of the following: 1. Trends 2. Changes

jobs. The more indistinct and overlapping the nature of a task, the more one needs to resort to indirect assessment of aggregate tasks and functions. All such assessments can be adequately performed using regression analysis.

In the case of direct jobs, for instance, regression analysis is used to generalize results that are obtained by direct observation or study. In the case of jobs that have a higher degree of complexity and ambiguity, but that do not lie at the right end of the spectrum, regression analysis is used to assess the relationship of the amount of time required to perform a task to different measures of task descriptors. At the extreme right end of the scale that corresponds to indirect labor, linear regression analysis is used to quantify the labor content of aggregate tasks and functions. This is accomplished by ascertaining the historical relationships of staff group sizes to major task indices.

For example, consider an assignment that requires you to ascertain how many personnel are required to perform the task of plant and equipment development engineering. One way to approach this assignment is to look at past data. Historical records of development expenses and past research on the topic might be closely related to product and process development staffs.

It is clear from the preceding discussion that job content is the least definable in the case of most white-collar jobs. At this end of the spectrum, employees are able to independently achieve efficiency and effectiveness in order to be productive. The natural behavior of such employees is to induce growth. Thus the sole task of management in such situations is to make aggregate judgments concerning validity. An answer is sought to the following question: "What level of indirect employment provides an optimal balance given the many conflicting constraints that might exist?" Some of the constraints referred to in the above question are listed here:

- Smooth assimilation of new staff members. Concern for recruitment, training and orientation, and provision of facilities.
- Avoidance of an extreme reaction to pressures concerning growth. This is important because an extreme negative reaction in such cases can lead to instability and inefficiency resulting in expensive turnover.
- Selecting projects that are cost-effective. Submarginal projects need to be avoided. Such projects are defined as those that offer stimulation and fun but are typically low yielding.

Thus, the goal of management is to be neither understaffed (thus missing important opportunities) nor be overstaffed (engaging in submarginal activities). The following examples depict application of multiple regression analysis to white-collar jobs or jobs that incorporate indirect labor:

- Programming time for CNC machining operations. This time is expressed as a function of machining specifications.
- The staffing requirements in a regional warehousing facility can be expressed as a function of various activities that require "counting" as an indispensable function.
- The amount of time required to set up a punch can be expressed as a function of die characteristics.
- Sales volume per dealer can be expressed as a function of characteristics of the market.
- Sales volume can also be expressed as a function of sales strategies adopted by sales staff. Not all sales strategies are effective. Therefore, some strategies that are experimental in nature can also be studied.
- Staffing requirements to perform a maintenance operation as well as time required to perform a maintenance operation can be expressed as a function of product characteristics and downtime that is not included in the schedule.
- Warehouse labor can be expressed as a function of various activities such as order filling and packing.

The greatest obstacle to modeling white-collar work is the shortage of aggregate data. Research or capital budgets can be regarded as examples of useful metrics, each of which are allocated on an annual basis. Historical data may go as far back as 15–20 years. Indexes of such functions are often confounded adding to the level of task complexity.

The relationship between predictor variables is subject to change over time. For example, effectiveness of some engineers with respect to their primary task may actually decrease over time due to the need to focus more on secondary tasks such as environmental conservation or legal ramifications of their projects.

Similarly, relationships being discussed in this section can be dispersed over time with some team members being involved in planning, staffing, and execution.

In general, the following principles need to be adhered to when performing regression modeling:

1. At the outset, reasonably specific goals need to be set for the modeling process.
2. Knowledge of the job type spectral domain is important because it helps in choosing modeling goals appropriate to that domain.
3. The client, system experts, and an analyst need to be formally included on a "project team."
4. Regression analysis involves a lot of modeling and analysis. Thus an appropriate computer program needs to be chosen that is capable of performing all the required computations.
5. Resort to a cyclical diagnostic and evolutionary approach to modeling that relies on the interaction between the system expert and the analyst.
6. Start with a basic preliminary data sample. The data can be extended as and how it is needed in order to enhance precision.
7. Models need to be built that reflect the complexity of the system. Focus on adequacy as the goal instead of simplicity.
8. Always remember the importance of an intercept term, the possibility of interactions as well as the utility of attribute coding.
9. The model adequacy always needs to be examined by asking appropriate questions. The answers to such questions can be obtained by using appropriate statistics.
10. Changes in the system are always to be expected; therefore, the model needs to be constantly updated and revised, even if the process is incremental in nature. Model maintenance is very important.

8.8 FUNCTIONAL PERFORMANCE SYSTEM

The functional performance system is also referred to as the common staffing study or the common staffing system. It was developed by IBM to measure indirect labor staffing requirements. It uses different productivity type measures in order to approximately measure groupings of indirect labor. Productivity type measures relate the amount of worker input required to produce measurable output.

This technique identifies discrete activities or tasks performed as an integral part of a specific job function. It will be appreciated that each indirect labor function with any organization is composed of smaller activities. An example is presented in Table 8.6 that subdivides administrative and personnel functions into individual activities. Each activity is performed by a set of individuals. They are associated with the number of employees performing them, commonly referred

Table 8.6: Division of Administration and Personnel Functions into Functional Activities

Code	Administration Tasks Title
101	Order entry
102	Administrative marketing support
103	Backlog management
104	Customer calls
105	Accounts receivable
110	Sales administration
111	DP administration programs
112	Scheduling
113	Inventory control
114	Field administration support
121	Telecommunications
122	Reprographics, office equipment, and supplies
123	General administration services
124	Security and safety
125	Cafeteria services
126	Mail collection and distribution
	Personnel Tasks
201	Personnel planning
202	Employment and placement
203	Personnel systems and records
204	Management development
205	Personnel and employee programs
206	Compensation
207	International assignments program
208	Benefits, reimbursements, and welfare programs
209	Suggestions, awards
210	Employee development and education

to as the headcount. Quantifiable cause and the activity headcount constitute a database from which performance indices are computed. Such indices can then be used as a measure of comparison between alternate staffing and productivity scenarios. Performance indices can be computed by adhering to the following steps.

- The main job functions are identified and then subdivided into constituent activities. The accounting function can be subclassified into constituent activities like that of estimating or buying.
- It will be appreciated that each activity happens for a specific identifiable reason. For instance, the accounts payable activity "happens" in order to process receipts. The secretarial services activity "happens" in order to support plant and other related indirect personnel.
- The volume of work contained with each activity (the amount of activity) that should take place is directly proportional to the amount of underlying reason. The activity "vendor billing" is directly proportional to the number of invoices processed.

- There is a specific workload for each activity. The workload corresponds to the total equivalent employees performing that activity during the evaluation timeframe. This includes full-time and parttime workers, regular and nonregular workers, etc.
- Once the equivalent headcount for each activity during an evaluation period has been determined, it is separated in keeping with the appropriate and relevant numerical work cause for the period under consideration. This information is used to compute several pertinent ratios.

1. *Productivity ratio*: This ratio links the work cause for each location to the equivalent headcount for each activity. Thus, from Figure 8.3, for plant X, the productivity ratio can be defined as

$$\frac{\text{Number of secretaries (100)}}{\text{Indirect employee population (1000)}}$$

2. *Mean productivity ratio*: The mean productivity ratio corresponds to the total-weighted ratio for all plants. It is computed by dividing the sum of all activity workloads for all the plants in the survey by the sum of all work cause quantities for all plants. Thus, in this specific example, assuming that the mean productivity ratio for secretarial activity is 0.095, we have the following expression:

$$\frac{\text{Sum total of secretaries for all plants}}{\text{Sum total of indirect employee population of all plants}} = 0.095$$

3. *Work load*: Next, the index workload is computed for each location. This corresponds to the amount of workload that would be involved if the plant's productivity ratio was equal to the mean productivity ratios for all plants. This is arrived at by multiplying the work–cause quantity of the plant by the mean productivity ratio. Thus for the plant under consideration, the index work load value can be written as:

$$\text{Indirect employee population (1000)} \times \text{Mean productivity ratio (0.095)} = 95$$

4. *Norm deviation*: The norm deviation for each plant is defined as an indication of the potential improvement for that plant. Thus, for the situation under consideration, we have the following:

$$\text{Actual work load (100)} - \text{Index work load (95)} = 5$$

5. *Norm index*: The norm index for each location compares the performance of each individual location to the average of all locations adjusted for differences in work-cause values. Thus, in this case, we have

$$\frac{\text{Actual work load (100)}}{\text{Index work load (95)}} = 1.053$$

6. *Productivity index*: The productivity index for each location tracks the change of performance of that location over time. Assume that the plant under consideration staffs 100 secretaries during the current year, which

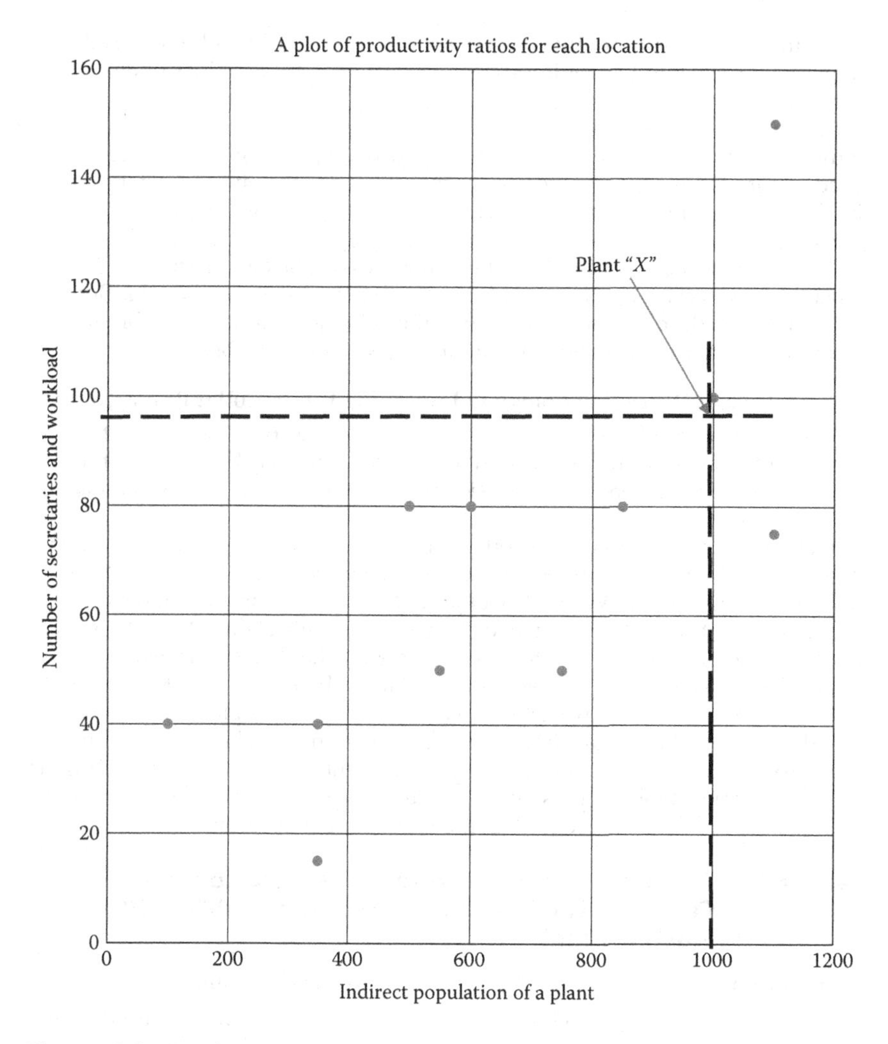

Figure 8.3 Productivity ratios by location of a plant.

is unchanged from the previous year. Also assume that the indirect employee population has risen from 800 last year to 1000 this year, then the productivity for such a scenario would be expressed as

$$\frac{\text{Productivity ratio for current year}}{\text{Productivity ratio for the preceding year}}$$

$$= \frac{\text{Secretaries in current year (100)} / \text{Indirect employee population in current year (1000)}}{\text{Secretaries in preceding year (100)} / \text{Indirect employee population in preceding year (800)}}$$

$$= \frac{0.1}{0.125} = 0.8$$

For individual locations, the norm index and productivity index are used to pinpoint opportunities for improvement relative to the activities under consideration.

As a rule of thumb, a norm index with a value greater than 1.0 tends to indicate a plant performing below the average of all plants and vice versa. A productivity index with a value greater than 1.0 tends to point to a deterioration in plant performance during time over which the study was conducted and vice versa.

It has been argued repeatedly that technology has played a crucial role in advancing overall productivity, especially in the case of blue-collar jobs. But this tends not to be the case with respect to white-collar jobs. Section 8.8.1 briefly examines the role of technology in enhancing white-collar work.

8.8.1 Role of Office Technology in Improving White-Collar Productivity

The cost associated with employing office-based white-collar workers in the United States amounts to billions of dollars. Approximately 73% of the total cost is attributed to managers and professionals while clerical staff account for about 27% of the cost.

Technology has replaced many office functions or so it seems. A closer examination indicates that information resources account for only about 9%–12% of total white-collar costs. As could be expected a large proportion, almost 70% of this expenditure, is attributable to clerical work. Only about 13% is directed toward professionals, and 17% toward managers. It has been observed that almost 25% of knowledge worker (white-collar worker) time is wasted by devotion to unproductive time wasters such as searching for people or information, scheduling, copying, or traveling to and from meetings. Technology aids in trying to fix this problem thus minimizing the amount of time wasted by indulging is such activities. Table 8.7 presents examples of typical time wasting activities accompanied by their respective technological fixes or solutions.

Table 8.7: A List of Possible Technological Solutions to Common Time Wasters (Account for 15%–40% of Wasted Time)

Time Wasters	Available Technological Fix
Seeking information on sources of supply	Online access to external supplier databases and internal records of supplier performance
Reaching key colleagues on the phone	An easy to use desktop keyboard or speech mail system
Traveling to periodic internal status meetings	Videoconferencing
Excessive "what-if" number crunching iterations	Automated decision support system
Seeking the status of an order	Desktop access to an information tracking system
Extensive corrections/revisions to documented reports and correspondence	Nearby access to a powerful word and graphics composition and revision processor
Scheduling meetings	Desktop displays of executive calendars available to both executives as well as secretaries
Generating reminders to subordinates to meet agreed upon schedules	Tickler system that generates automated reminders

Technology substantially eliminates, or at the very least meaningfully modifies, many time wasters. It also has the ability to allow better, well-informed decisions to be made, thus enhancing white-collar workers' contribution to the mission of the organization. This can indirectly enhance the quality of working life, which by and of itself contributes a great amount of intangible value. It is logical to conclude that the economic benefits of adopting office technology are substantial.

It has been observed that as much as a 9% increase in white-collar productivity can be achieved in about 18–24 months through the selective application of appropriate technology. This amounts to a rate of return on investment of more than 50%. Assuming that the management continues on its mission to enhance office productivity by resorting to selectively apply appropriate technology, a productivity gain of about 15% can be easily achieved over a period of 5 years and substantial benefits accruing over 10 years. In light of its obvious benefits, office technology needs to be managed judiciously. The following are the key elements of this management practice:

- Technology assessment and forecasting are essential to fully understand the current and forthcoming nature of office technology.
- Technology forecasting often takes the form of economic, demographic, and social forecasting in order to understand changes that are likely to happen and their effects on the organization.
- Delineation of the role of office technology, definition of its functional mission and objectives, position within the organization, authorities, responsibilities, and working relationships need to be clearly defined.
- Strategic plans need to be developed based on room for application throughout the organization, assessment of benefit–cost relationship, and prioritization of application to a specific job function and projects.
- Tactical plans need to be developed to enable applications identified in strategic planning, the amount and nature of resources required, developmental time, funding and need for coordination need to be carefully delineated.
- Strategic and tactical plans need to be revised with special emphasis on project execution and implementation.
- Office technology-related projects need to be executed with an emphasis on provision of relevant technical and organizational behavior support, ongoing project management, and significant user involvement.

8.9 SUMMARY

No book on work measurement can be complete without a discussion on white-collar work, its role in the modern organization and its measurement. This chapter dealt with the increasing role of white-collar workers or the so-called "knowledge workers" in the modern workplace. Different methods of measuring white-collar work were presented and appropriate examples were cited to highlight their utility value in the office. The ever-changing role of technology was also discussed in brief and the fact that office-based technology (white collar) is significantly different from blue-collar technology was emphasized.

9 Nontraditional Methods of Measuring Work

9.1 INTRODUCTION

The previous chapters in this book have focused on *time* as the measure of work in general, or a task in particular. Measuring the time it takes to perform a job and basing work standards on that, however, results in variations in work standards (time standards) across industry for the same job. This, as discussed before, is primarily due to subjectivity in rating the pace of performing a task, leading to different levels of "normal" within a firm, and the determination of allowances, particularly fatigue allowances. Rest allowances which often are an arbitrary percentage of erroneous normal times (also called basic times), therefore, are either inadequate or unnecessary. Moreover, the same rest allowance is applicable to each worker since individual differences are ignored.

In the event that rest allowances are excessive, organizational productivity is directly affected. In the event that rest allowances are inadequate, it leads to acute and chronic worker fatigue. This not only leads to reduced worker productivity but could also be the cause of injuries; overexertion injuries in industry are a major cause of concern. With each injury, certain costs are associated. For example, medical cost, lost work days cost, replacement cost, etc. The ultimate effect of all these costs is reduced organizational productivity.

In today's environment, there is a growing movement to automate work and work places. The underlying reason for this trend is the need to enhance productivity. Although the trend to mechanize and automate jobs has caused significant reduction in physical demands on workers, physical work remains a major part of many different occupations. Strong social, economic, and technological reasons are forcing retention of physical components in many industrial jobs. As we are getting closer to the limits of physical productivity, even small gains in productivity are becoming critical. Jobs that are poorly designed or that do not offer proper rest allowances cannot only offset these relatively small gains, but produce a large productivity deficit. The detrimental effect on productivity is due to undesirable levels of acute and chronic fatigue and due to the resulting injuries to workers. The costs of such job-related injuries are in fact a loss to the organization since the funds cannot be used for economic growth. The net effect is a decline in productivity.

Industrial engineers have always attempted to design jobs and workplaces for comfort, efficiency, safety, and, thereby, greater productivity. In Chapter 4, we looked at how fatigue allowances can be determined more reliably and objectively; for instance, the use of oxygen consumption or metabolic energy expenditure rate to determine variable fatigue allowances. The other option is to determine levels of work intensities that can be performed without causing excessive fatigue and overexertion. There are also many work situations where objective methods to measure work, such as those based on physiological evaluation, are the only appropriate means of determining acceptable levels of work and rest allowances; for instance, work performed in hot and humid climates or at high elevations.

In this chapter, we look at the nontraditional methods of measuring work, acceptable levels of work and, wherever appropriate, determination of fatigue rest allowances. The situations where these methods are used do not lend themselves to using *time* as an appropriate measure of work. The methods that are discussed are based on the following:

- Metabolic energy expenditure or oxygen consumption
- Heart rate
- Core body temperature
- Ratings of perceived exertion
- Critical fusion frequency
- Measurement of subjective feelings
- Psychomotor tests

The first three of the above methods can be classified as physiological methods and the remaining four can be classified as psychological methods. All these methods are widely used in scientific studies of work and work-related fatigue.

9.2 METABOLIC ENERGY EXPENDITURE OR OXYGEN CONSUMPTION

Physical work requires muscular exertions, which in turn require energy. This energy is provided by burning of the fuel (food) that people intake. Burning of fuel requires oxygen. If we can measure oxygen that is consumed while performing a task, then the energy requirements of the task can be determined. *It should be noted that this energy requirement not only includes the energy required by the task but also by the body.* Each liter of oxygen consumed provides 5 kcal of energy (=20.92 kJ; each kcal is equivalent to 4.184 kJ of energy). The amount of oxygen is measured by measuring the volume of inspired air times the percentage of oxygen in the intake air (the amount of oxygen in inspired air varies between 20.90% and 20.94%, depending on the moisture level in the inspired air) less the volume of expired air times the percentage of oxygen in the expired air. Sophisticated portable instruments are available to calculate and provide the amount of oxygen consumed per minute by an individual performing an activity (some of the instruments are discussed later on in this section). Thus, the energy expended on a job is determined by measuring the volume of oxygen consumed while performing that job. The energy expended on the job integrates factors such as the type of job, body posture assumed by the worker, magnitude of force exertion, age, gender, etc.

The beginning of work leads to an increase in the demand for energy. This increase in energy requirement, as indicated by increase in oxygen consumption, after a few minutes of work reaches a "steady state." As shown in Figure 9.1, oxygen consumption, and thus the liberation of metabolic energy by the body increases slowly due to the sluggish response of the respiratory and circulatory systems. The supply of energy during the initial period of work is mainly by the anaerobic process (energy liberation in the absence of oxygen). When the supply of oxygen is able to cope up with the task demand, the process becomes aerobic (energy liberation using oxygen in the inhaled air). During the initial stages of work, anaerobic process mostly supplies the needed energy, and lactic acid accumulates in the body. Some lactic acid also continues to accumulate in insignificant amounts even after the supply of oxygen reaches a steady state (indicated by the flat part in Figure 9.1). This lactic acid buildup must be eliminated and some oxygen is, therefore, required for that purpose when the work stops (imagine running—one continues to huff and puff even after running ceases).

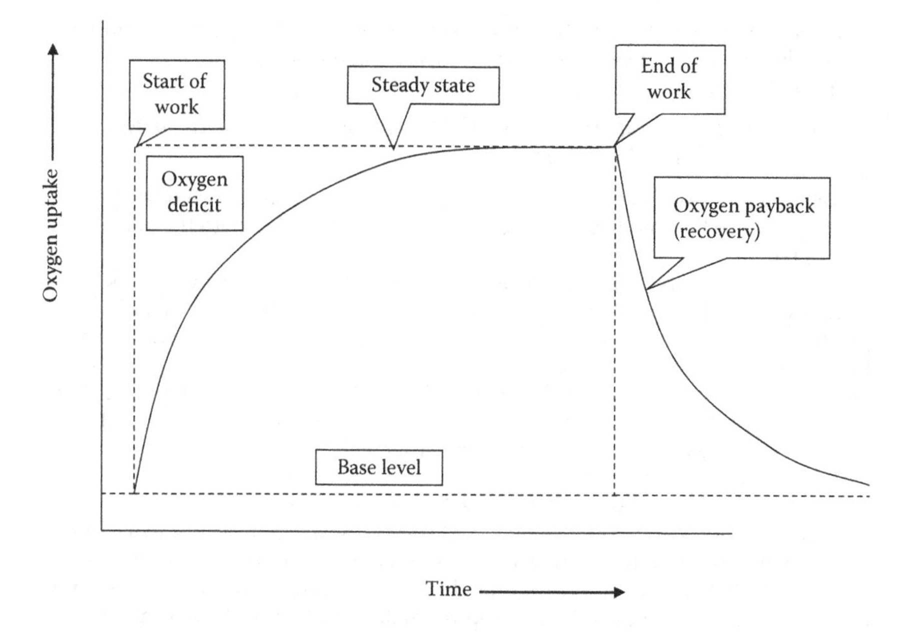

Figure 9.1 Increase in oxygen uptake with time.

Table 9.1: Influence of Selected Work and Workplace Factors Affecting Metabolic Energy Requirement

Factor	Net Effect of Metabolic Energy Requirement
Pace of work ↑	Increase
Task duration ↕	Increase or decrease depending upon work
Intensity ↑	Increase
Force ↑	Increase
Distance traveled ↑	Increase
Gradient ↑	Increase
Speed ↑	Increase

Table 9.1 shows some of the work and workplace factors that influence metabolic energy requirements. Additionally, some personal and environmental factors influence metabolic energy requirements, and hence oxygen consumption, as well. At this stage, we have two basic approaches to deal with the establishing work standards:

1. Develop work standards based on metabolic energy requirements that can be sustained without undue fatigue, thus negating the need for variable fatigue allowances.
2. Determine variable fatigue allowances that would be needed based on the metabolic energy needs.

The first approach requires knowledge of what are the sustainable levels of metabolic energy expenditure rates (kcal/min) that can be sustained without causing undue fatigue. Scientific studies have shown that

- *For an 8-h work shift, whole body tasks that require 28%–29% of aerobic capacity (see the subsection on measuring aerobic capacity in this section) can be sustained without undue fatigue.*
- *For 12 h shifts, physical tasks requiring 23%–24% of aerobic capacity can be sustained without undue fatigue.*

For shift durations between 8 and 12 h, interpolation may be used (extrapolation may also be used if the shift durations are not too far from the above numbers).

Even though we have provided recommendations for establishing work standards based on what workloads can be sustained without undue fatigue, production, technical, and economic factors frequently dictate that workloads be higher. In such cases, the second approach, that is determining variable fatigue allowances, needs to be utilized. Section 4.3 describes how the metabolic energy expenditure-based fatigue allowances can be determined. Appendix lists the software that may be used for such purpose as well.

9.2.1 Measuring Aerobic Capacity

Aerobic capacity, also known as maximal aerobic power or physical work capacity, is the maximum rate of oxygen uptake of a person. For male nonathletes, the average aerobic capacity is approximately 3 L of oxygen per minute. For female nonathletes, the aerobic capacity is approximately 2.6 L/min. This means that the maximum rate at which industrious men can expend energy is 15 kcal/min; for industrious women, this number would be 13 kcal/min. Of course, these levels of metabolic energy expenditure rates can only be sustained for few short minutes. As recommended above, for prolonged periods the metabolic energy expenditure rate has to be much lower if the buildup of fatigue is to be avoided. *This means that if fatigue is to be avoided, work rates for male workers should have metabolic energy demands that do not exceed 4.3 kcal/min; the corresponding level for women would be approximately 3.8 kcal/min.* These numbers are for the population. To be more precise, we need to know what the actual aerobic capacity of a worker is (it should be noted that developing work standards based on individual capacities is neither desirable nor practical as jobs are performed by a variety of individuals, men and women).

Several methods have been described in the published literature for measuring aerobic capacity. Here, we describe a simple method for measuring aerobic capacity using a bicycle ergometer. The method does not require that individuals be subjected to maximal levels of work rates in order to determine aerobic capacity. The submaximal bicycle ergometer method of determining aerobic capacity requires that an individual be subjected to three different levels of work rate (the typical levels of work rate are 100, 150, and 200 W), and the amount of oxygen that is consumed at each work rate is recorded once the steady state is reached (see Figure 9.1). The heart rate at these work rates is also recorded once the steady state is reached. The heart rate and oxygen uptake at each work rate are plotted and a best-fit straight line is drawn through the plotted points (see Figure 9.2). Once the straight line is plotted, oxygen uptake corresponding to the *maximum heart rate* is determined as shown in Figure 9.2. The maximum heart rate is determined from the following equation:

$$\text{Maximum heart rate (MHR)} = 214 - 0.71 \times \text{Age in years}$$

The oxygen uptake corresponding to the maximum heart rate is read and is the *aerobic capacity* of the individual who is subjected to the testing.

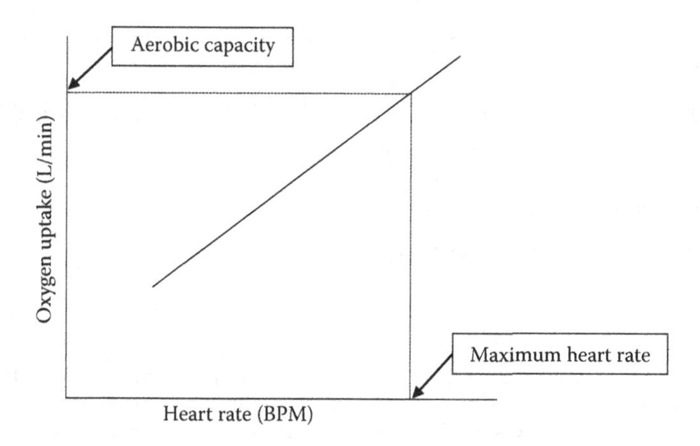

Figure 9.2 Aerobic capacity determination using submaximal bicycle ergometer test.

Consider 28%–29% of the aerobic capacity would provide metabolic energy expenditure rate, and hence the work rate, that can be sustained for an 8-h work shift without fatigue (23%–24% for 12-h shifts).

As previously stated, using work rates based on an individual's aerobic capacity as work standards is neither desirable not practical. Jobs should permit using a wide variety of workers, both men and women. Therefore, the numbers suggested earlier for the population should be used (4.3 kcal/min for men and 3.8 kcal/min for women; if both men and women perform the job, a lower number should be used). Since this approach may not work for production, technical, and economic reasons, it is recommended that rest allowances be determined for actual work rates as suggested in Section 4.3. *The key with physiological fatigue allowances is to remember that breaks should be short and frequent, increasing in duration towards the end of the shift. This is because the buildup of fatigue, as well as recovery from fatigue, is exponential. As much as 95% of recovery takes place in the first few minutes of the break. The remaining 5% of the fatigue takes the bulk of the rest break. Fatigue buildup continues throughout the shift even though rest breaks are given throughout. It is for that reason that rest breaks should increase in duration as the shift progresses.*

At high altitudes, 2000–4500 m, aerobic capacity can decline by as much as 15%–30%. Therefore, the guidelines need to account for this decline particularly if the individual is not acclimatized. Even acclimatized individuals may have a reduction of as much as 10% in aerobic capacity. Workloads based on aerobic capacity, therefore, are important rather than just using flat numbers such as 4.3 kcal/min or 3.8 kcal/min.

9.2.2 Instruments for Measuring Oxygen Uptake

Direct measurement of oxygen consumption at the job is important. While a variety of oxygen measuring equipment are available, not all equipment are suitable for use on the shop floor. Portable equipment is suitable because it can be taken from one place to another. A portable oxygen uptake measuring equipment should have the following attributes:

- It should be light
- It should be battery operated
- It should be capable of measuring the volume of inspired and expired air

- It should be capable of measuring oxygen concentration in inspired and expired air
- It should be able to calculate oxygen uptake rapidly (using a microprocessor)
- It should provide a printout of oxygen uptake if needed

The following three systems possess these attributes:

- TEEM 100 (total energy expenditure measurement system)
- COSMED K2 system
- Oxylog

The COSMED system is a telemetric system, equipped with an FM radio transmitter that broadcasts its signal to a receiver. It works for distances up to 600 m. This feature allows workers to work without intrusion from observers and makes the system very attractive.

The choice of a system would primarily be based on the cost, as all these systems are reliable and fairly accurate. While TEEM 100 and Oxylog are priced under $10,000, the COSMED K2 costs more than $35,000 due to its telemetry feature. Therefore, one needs to ascertain if telemetry is really needed.

9.3 HEART RATE

While the measurement of metabolic energy expenditure provides a more objective and accurate assessment of fatigue and workloads, it is expensive because of the cost of the equipment. Since it also requires the use of a mask during work, some might consider that intrusive. Measurement of heart rate provides an alternative as it increases with the metabolic energy expenditure and, as shown in Figure 9.2, there is a linear relationship between the two. The following need to be considered:

- Heart rate varies considerably from individual to individual
- Heart rate as a measure is suitable only for dynamic work involving whole body
- Heart rate is suitable for work with steady rhythm
- Heart rate is affected by age and physical conditioning of the individual
- Heart rate is significantly influenced by heat and humidity
- Heart rate is susceptible to emotional or psychological stresses

To use heart rate as a measure of workload and fatigue, influence of these factors must be minimized or restricted.

When work is light or moderate, heart rate increases to levels appropriate to the physical effort and then stabilizes. It remains constant thereafter for the duration of the work. At the end of the work, it returns to normal (heart rate at resting level). If, on the other hand, work is strenuous, heart rate continues to increase and never stabilizes, until the worker is exhausted. Figure 9.3 shows changes in heart rate for different workloads.

In Figure 9.3:

- Resting pulse is the average heart rate before the work begins
- Working pulse is the average heart rate during work

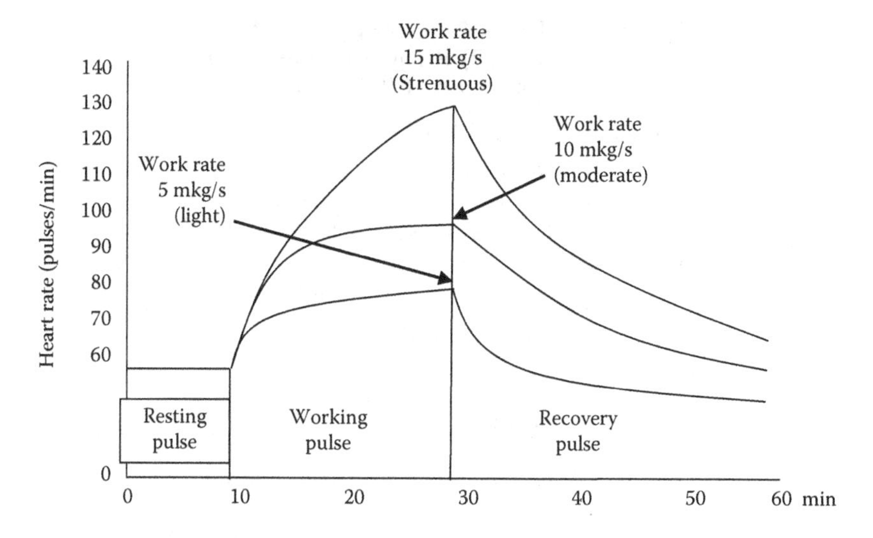

Figure 9.3 Changes in heart rate for different workloads.

- Recovery pulse is the heart rate after the work stops
- Total recovery pulse is the sum of heartbeats from work stoppage until the pulse returns to resting level

Based on heart rate, acceptable upper limit of work is defined as workload for which the working pulse does not continue to rise, achieving a constant level (unlike the case for strenuous work in Figure 9.3*), and returns to resting level 15 min after the work stops.* This limit tends to ensure that the metabolic energy is being used at the same rate as it is being replaced and a steady state can be maintained.

The limit for continuous work performance for men is a working pulse that is 35 beats/min above the resting pulse. For women, the limit for continuous work is a working pulse that is 30 beats/min above the resting pulse. The resting pulse is measured while the worker is seated. These heart rate limits correspond well with the metabolic energy expenditure-based limits as 10 work pulses are shown to correspond to metabolic energy expenditure levels of 4 kJ/min.

As in the case of metabolic energy expenditure, heart rate tends to rise gradually with the working duration due to the accumulation of fatigue (accumulation of lactic acid). This is called heart rate creep. It is important that heart rate be monitored carefully when the workloads are being set.

In the event, workloads cannot be set to avoid fatigue, rest allowances may be determined as discussed in Section 4.3.

The equipment for measuring heart rate is simple. One of the most common types of heart rate monitors is a watch that can be worn on the wrist. It is reliable and accurate. The cost of such equipment is well under $500.

9.4 RATINGS OF PERCEIVED EXERTION

While heart rate is a simpler measure of workloads in comparison to metabolic energy expenditure rate, an even simpler measure of physical exertion is the *ratings of perceived exertion* (RPE). It requires no equipment and is fairly reliable. The scale that allows workers to subjectively assess and rate physical exertion

Table 9.2: Borg Category Ratio 10 (CR10) Rating of Perceived Exertion (RPE) Scale

Rating	Verbal Anchor	
0	Nothing at all	Resting
0.3		
0.5	Extremely weak	Just noticeable
0.7		
1	Very weak	Really easy
1.5		
2	Weak	Easy
2.5		
3	Moderate	
4		
5	Strong[a]	Heavy or hard
6		
7	Very strong	Really hard
8		
9		
10	Extremely strong[b]	Maximal
11		
•	Absolute maximum	Highest possible

[a] 50% of aerobic capacity.
[b] Aerobic capacity.

and fatigue involved in manual work is known as *Borg CR10 scale*. This scale is a 10-point scale and is shown in Table 9.2. The verbal anchors associated with ratings are also shown.

The ratings of the Borg CR10 scale grow linearly with heart rate and metabolic energy expenditure rate and allow workers to perceive physical effort involved in workload intensity and work duration combinations. A 10 rating (maximal) on Borg CR10 scale would correspond to metabolic energy requirements equivalent to aerobic capacity (maximum oxygen uptake). A 5 rating (strong) on Borg CR10 scale would correspond to metabolic energy requirements equivalent to 50% of aerobic capacity. The Borg CR10 scale is an improvement over the popularly used Borg RPE scale with ratings from 6 to 20, and is specially designed to assess fatigue and pain and discomfort during musculoskeletal efforts. Since the ratings of the Borg CR10 scale are linearly related to heart rate and metabolic energy expenditure rate, acceptable levels of workloads should not elicit a rating exceeding 3–4 for continuous work.

Since the RPE ratings are subjective and can be influenced by prior experience, motivation, and other personal factors, one of the two approaches may be adopted:

1. Normalize ratings to each individual's maximum
2. Use values averaged over the ratings of a group of workers

The second approach is preferable since a job may have a variety of workers performing it. Further, this approach also accounts for personal factors such as gender, physical condition, and age.

9.5 BODY CORE TEMPERATURE

The normal body core temperature fluctuates narrowly around a normal of 98.6°F (37°C). The body gains heat or loses heat depending on the surrounding environmental conditions. The overall heat exchange between the body and the surrounding environment is given by

$$S = M \pm C \pm R - E$$

where M = Metabolic heat gain
C = Heat exchange through convection
R = Heat exchange through radiation
E = Heat loss through evaporation
S = Net heat loss or gain

As long as the net heat gain or loss, S, remains zero, the body remains in a state of thermal equilibrium. This equilibrium is known as *homeostasis*. The body core temperature is maintained at normal level during homeostasis. The process that regulates and maintains the body core temperature is called thermoregulation. Thermoregulation allows the body metabolic heat and the heat gained from the surroundings to be dissipated through convection and radiation. When convection and radiation are inadequate, body starts sweating and the accumulated heat is lost through evaporation of the sweat. If all these processes are unable to regulate the core temperature, then the core temperature starts rising. This rise in the core temperature, if significant, leads to a decline in performance. The National Institute of Occupational Safety and Health (NIOSH) in the United States has established that the maximum allowable heat gain be limited to an increase of 1°C (about 1.8°F) in core temperature. Workloads and heat gain from the environment can be sustained and work can continue without interruption as long as the core temperature does not exceed 38°C. Continual increase in core temperature can lead to heat stress, heat exhaustion, heat stroke and, eventually, death.

Excessive heat is one stressor to which most workers are exposed to at one time or another. Exposure to heat is pervasive in some vocations, such as foundries and steel mills. There are two options to limit worker exposure to heat stress:

- Modify the environment
- Limit worker exposure to heat by providing rest allowances

The modification of environment requires that metabolic load, and convective and radiative heat gains be reduced. Reducing the workload can reduce the metabolic heat; increasing airflow, reducing humidity by using dehumidifiers, and reducing the surrounding air temperature can reduce convective heat gain; and radiative heat gain can be reduced by shielding the worker from surrounding heat sources. Additionally, evaporative heat loss from the worker can be increased by exposing skin, provided it does not cause radiative heat gain, and by reducing humidity.

If modifications to the environment are unable to control heat stress, worker exposure to heat must be limited. Section 4.3 provides a relationship for determining rest allowances as a function of metabolic load and the wet bulb globe temperature (WBGT). This relationship is based on the NIOSH guideline that the maximum increase in core temperature must be limited to 1°C. Keeping this recommendation in mind, Figure 9.4 shows the limits of metabolic workloads and environmental conditions for continuous work (solid line). It also shows

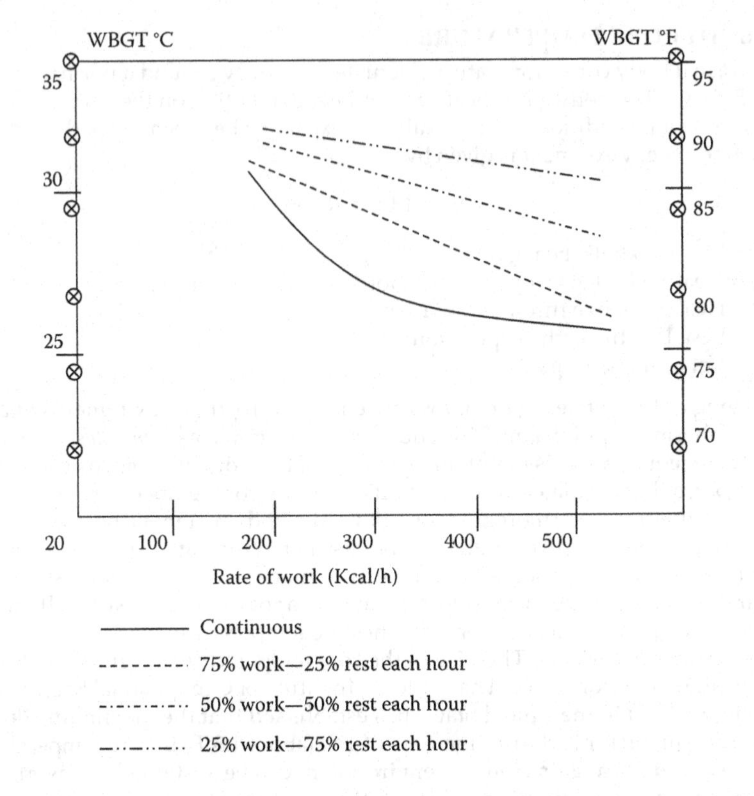

Figure 9.4 Limits of heat exposure.

different rest allowance–work duration combinations for various workloads and environmental conditions. One can use either the relationship provided in Section 4.3 or Figure 9.4 to determine rest allowances and working durations for different heat conditions. The work standards that allow continuous work and conform to NIOSH guideline regarding the core temperature can also be obtained from Figure 9.4.

9.6 FLICKER FUSION FREQUENCY

Flicker fusion frequency, sometimes also referred to as critical fusion frequency, is the frequency at which the flickers of a flickering light source (lamp) appear to fuse into a continuous light. A reduction of 0.5–6 Hz in FFF takes place under mental stress and various other industrial stresses, such as visual stress; not every type of stress leads to a reduction in FFF. The lowering of FFF is more pronounced during high level of stress and is often accompanied by other signs of fatigue, such as increased feelings of weariness and sleepiness. The FFF is measured by increasing the frequency of flickers until the flickers fuse into a continuous light. The source of the light should subtend an angle of 1–2° at the eye and should not require any visual accommodation.

FFF is affected by the intensity of light source and whether the vision is mediated by rod vision (black and white or night vision) or cone vision (color vision). For rod-mediated vision, normal FFF may be about 15 Hz, whereas for cone-mediated vision, FFF may be as high as 60 Hz.

Generally, there is a good correlation between lowering of FFF and subjective estimates of fatigue. Since FFF is affected by age and some medications, it is a good idea to establish the normal level of FFF for an individual before using it as a measure of fatigue and for work standards. *Work standards may be based on stress levels that do not cause a reduction in FFF and show no presence of subjective feelings of fatigue.* If there is a significant decline in FFF, one can reduce the stress level of the task to a moderate level of mental effort so that work may continue without interruptions, or provide rest allowances. The method described in Section 4.3 for mental fatigue may be used for this purpose.

9.7 SUBJECTIVE FEELINGS

Bipolar questionnaires can be used to assess subjective feelings of fatigue. These questionnaires could be simple or complex. The following is an example of a simple bipolar questionnaire showing opposing items.

Fresh ____________________________ Weary

Sleepy ____________________________ Alert

Weak ____________________________ Strong

Sick ____________________________ Healthy

Optimistic ____________________________ Pessimistic

Keen ____________________________ Bored

Attentive ____________________________ Absent minded

Energetic ____________________________ Exhausted

Ready ____________________________ Indifferent

Even though the bipolar questionnaires are simple, they do provide an indication of a worker's subjective feelings of fatigue. Workloads and work durations that indicate the onset of fatigue need to have rest allowances. It is a good idea if a bipolar questionnaire is accompanied by a mental test involving arithmetic problems. The outcome can then be used in the methods described in Section 4.3 for assessing rest allowances.

It is recommended that occupations that require high levels of mental stress or sustained levels of vigilance should have planned work and rest periods in order to avoid unacceptable levels of errors and accidents.

9.8 SUMMARY

In this chapter, we have reviewed some methods that can allow measurement of work when time/unit may not be a suitable measure. Some of these methods are physiological in nature and some are clearly psychological and subjective. Physiological methods are objective and more accurate as compared with psychological methods. All methods, however, put emphasis on individual performance and most lead to work standards for individual workers rather than a group of workers. One way to develop group standards would be to look at individual performances and then use statistical techniques (e.g., use of normal distribution) to develop work standards that will be suitable for a specified population (e.g., upper 90% or 95%). These statistical techniques are not discussed here but any text on statistics could provide such basic information.

10 Instrumentation and Software

10.1 INTRODUCTION

This book provides a quick overview of work-measurement-related concepts for all engineers, not just IEs. The material presented in this book so far deals with basic concepts and their applications. However, rather cursory information has been provided with respect to these "tools of the trade."

Work measurement analysts routinely use different types of instrumentation to measure work. The accompanying software is used to examine the results of work measurement. To fully appreciate the scope and nature of the task, it is crucial to gain familiarity of relevant instrumentation and software. This is especially important in light of the higher level of sophistication and precision afforded by continuously evolving technological advances in electronics and software. For instance, stopwatches were used for the first time in the 1940s to conduct time studies. In keeping with the level of technological sophistication available at the time, mechanical movements largely powered such watches. The level of precision available to the work measurement analysts was also limited and was expressed as a function of the least count of the measuring instrument. Most analysis of observations was conducted manually using a paper and pencil. The computational power available to analysts was almost nonexistent. This is no longer the case. For instance, analysts routinely use spreadsheets to perform complicated calculations and present the results, mostly in easily recognizable graphical form. The advent of personal computers, advances in computational technology, and miniaturization have radically altered the method of conducting the "trade" and analyzing the results. Thus, it is essential for the modern work measurement analyst to be fully cognizant of the vastly improved tools and capability available to them in order to be fully effective.

This chapter seeks to do just that. Examples of instrumentation will be presented along with images of the "instruments" under consideration. The outstanding technical features of each instrument, its area of specific application as well as its cost will also be discussed.

Discussion of instrumentation and software will be presented in the order in which they first appear in this book. Thus, stopwatches, for instance, will be presented when they were first discussed in the context of stopwatch time study. The sequence will be adhered to throughout the course of this chapter.

10.2 STOPWATCHES

Time study was discussed at length in Chapter 3. Some basic concepts were also alluded to in Chapter 2. It is essential to perform accurate time studies in order to effectively measure work and it constitutes the basis of all work measurement. Concepts such as PMTS, standard data, etc. have all evolved from fundamental time studies. Time has always been measured using a clock or a watch. In the context of work measurement, a stopwatch is used to accomplish this task.

Stopwatches have evolved considerably over time. Although mechanical stopwatches are still available, it is not mandatory by any means to use a mechanical stopwatch to perform a time study. Contemporary instruments of time measurement are electronic stopwatches. An electronic stopwatch offers several advantages over its mechanical counterpart. Some of these advantages are enumerated as follows:

- Greater accuracy
- Requires less maintenance and higher robustness

- Offers several different modes of time measurement
- Less expensive due to use of electronic circuits

Some examples of stopwatches (mechanical and electronic) are presented in the following pages:

Mechanical stopwatch: Mechanical stopwatches constitute the traditional instrument used for measuring work. The analog display is used to calibrate a mechanical stopwatch. The least count of a mechanical stopwatch can be calibrated in a range from about one-hundredth of one minute to one-fifth of one second. It can run for about 60 min. It is useful to perform continuous time studies. However, a snapback arrangement is available on most stopwatches, thus broadening their usability to include snapback time studies as well. An example of a mechanical stopwatch is depicted in Figure 10.1. Note the mechanical snapback arrangement on top of the dial. Mechanical stopwatches are available staring at about $40 and can run anywhere up to $300 depending on their precision, build quality, the quality of materials used, fit, finish, and reputability of the brand.

Electronic stopwatch: Most electronic stopwatches are operated by a battery and use the digital LCD mode of display (Figure 10.2). This is unlike the mechanical stopwatch that uses analog display. The accuracy of an electronic stopwatch is generally greater than its mechanical counterpart. Moreover, it is easier to read a digital display. The time measurement analyst's work is simplified in that the only task that needs to be performed correctly is to start and stop the watch at the right instant in time. The amount of time lapsed can be easily and accurately read from the digital display. The electronic stopwatch has reset functions similar to a mechanical watch. Memory constitutes another useful feature built into the electronic stopwatch. Thus previous times can be recalled if the need arises.

Figure 10.1 Mechanical stopwatch.

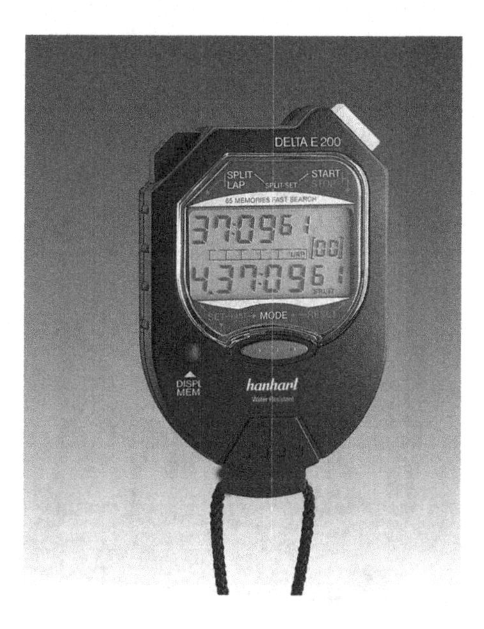

Figure 10.2 Electronic stopwatch.

The least count of such a watch can be as low as one-hundredth of one second. It can range in price from about $5 to $100 depending on the level of sophistication. Unlike its mechanical counterpart, the electronic stopwatch requires very little routine maintenance if any at all.

Tabletop stopwatch: Tabletop stopwatches share many of the features as the conventional electronic stopwatch. The only difference lies in the area of specific application. These watches are bigger, less portable, have larger displays, and are used in a sedentary setting. Figure 10.3 depicts a tabletop stopwatch. The least count can range from one-hundredth of a second to one-hundredth of a minute and prices can vary from $30 to $140.

Stopwatch boards: Performing a time study is tedious if the observer does not have access to the right equipment. All time studies that require observation of a task and recording time for individual activities involve at least two types of equipment: 1. A stopwatch and 2. A piece of paper to record details of the study in real time. Carrying a separate stopwatch along with a piece of paper can be quite cumbersome. To resolve this situation, Meylan Corp introduced a time study board that is available in various configurations (quick click system), one of which is depicted in Figure 10.4. The equipment consists of a writing pad with three stopwatches mounted on top. The following sequence of operations describes the manner in which the board can be used:

1. Set first watch at zero
2. Start and stop second watch
3. Have the third watch in motion
4. First quick click of the lever starts timer no. 1, clocks no. 2, and stops no. 3
5. At the end of first element, press quick click lever to stop no. 1 while no. 2 starts simultaneously

Figure 10.3 Tabletop stopwatch.

6. During this time, clock no. 1 is waiting to be read at the operator's convenience while the element is being timed by clock no. 2
7. Clock no. 3 has automatically reset at this point in time and is awaiting the third element
8. The automatic action described herein continues indefinitely with each quick click of the lever until all elements of the cycle have been timed

Figure 10.4 is composed of two pictures:

1. The first picture depicts the time study board in its entirety.
2. The second picture depicts the three stopwatches mounted on top of the board in detail. The three watches are exactly identical. Note the reset lever as well as the level of detail available in each watch. It needs to be mentioned that not all time study boards are equipped with analog stopwatches. Digital watches are being used with increasing frequency in order to facilitate the process of reading the time, thus minimizing analyst error.

10.3 TIME STUDY SOFTWARE

Time study is somewhat of a misnomer. It creates the assumption that the process consists of recording time for an activity (work) and then leaving it alone. We have seen throughout this book that this is certainly not the case (especially Chapters 3 and 6). The process of capturing activity time data is only an initial step. It needs to be analyzed in detail in order to be used correctly.

This book has presented numerous analysis techniques, each of which is quite rigorous in its use of mathematical principles. It will be appreciated that the analysis of time study data is a task that requires a high level of skill and dexterity. Given the complex nature of mathematical calculations and the high

(a)

(b)

Figure 10.4 Three-watch time study board with details.

volume of raw data often available to the analyst, it is quite common to encounter frequent mistakes. To solve this problem, numerous software packages have been available on the market. They perform a variety of functions ranging from mathematical calculations of raw data to presentation of data in a graphical format. In these pages, we will discuss a few software packages that deal with time and motion study analysis.

10.3.1 Proplanner

Proplanner is a company based in Iowa that offers process engineering and management software for manufacturers involved in product assembly. The applications such as automate, streamline, and integrate engineering activities are offered by the company. The software package offered by Proplanner (www.proplanner.com) contains seven main modules that are detailed below:

1. *Assembly planner*: Assembly planner is a suite of tools that address the manufacturing processes by linking to a database in order to help engineers create and manage prices information to improve manufacturing operations.
2. *Shop floor viewer*: Shop floor viewer gives shop floor workers direct access to view electronic work instructions through a web browser.
3. *Logistics database*: This is a materials handling database and contains a company's internal logistics and materials handling information.
4. *ProTime estimation*: This tool is directly in keeping with the content of this book. It provides an easy to use spreadsheet interface that enables manipulation of observed tasks (time study) or predetermined elements. Additionally, it also computes personal, fatigue, and delay allowances as well as any additional user-defined allowances. It supports most industry standards such as MTM-1, MTM-2, MTM-UAS, MTM-B, MTM-SAM, MTM-MEK, Modapts, and MOST (described in Chapter 6).
5. *ProBalance*: This tool offers an easy-to-use spreadsheet interface to manipulate tasks, stations, tools, and precedence. It provides sophisticated charting and line balancing techniques and machine/equipment utilization computation.
6. *Flow planner*: This tool is used to aid in facilities layout and ensures that facility layout changes are actual improvements. It interfaces with AutoCAD to generate material flow diagrams and cost of a layout generated in AutoCAD.
7. *Workplace planner*: This tool interfaces with AutoCAD and computes worker walk times, travel distances, and lean value-added percentages. Some relevant screen shots of the software are presented in the Figures 10.5 through 10.12.

10.3.2 Timer Pro Professional

Time Pro Professional is a company that also offers a suite of solutions for time and motion study applications as well as line balancing, video-based lean, Kaizen, and training. It offers several tools (listed on its website), some of which are listed below:

- Time and motion study
- Video time and motion
- Value-added analysis

Page 1 of 1

Time Study Report

ID: Sample Study
Time Type: Calculated
Total Time: 134.93 Seconds
Report On: 07/20/2011
Report By: PROPLANNER2 \christina
Modified On: 07/20/2011
Modified By: PROPLANNER2 \christina

Description: Contains an observed, calculated, and estimated time study for a single process.

Task Summary

	ID	Seq	Description	Man/Mach/Misc	Time	Ext. Time	Allowance	Int	Freq	VA	SVA	NVA
1	4.01	1.00	Collect filter assemblies and place in stand	Machine	15.12	15.12	0.00		1	3.96	0.00	0.00
2	4.03	3.00	Turn in hand screws to secure assy to fixtures	Manual	5.04	0.00	0.00	1	1	0.00	0.00	5.04
3	4.02	2.00	Adjust alignment on pins	Manual	2.52	2.52	0.00		1	0.00	0.00	2.52
4	4.04	4.00	Remove caps	Miscellaneous	14.76	14.76	0.00		1	0.00	0.00	14.76
5	4.05	5.00	Collect connector fittings	Manual	3.96	3.96	0.00		1	0.00	0.00	3.96
6	4.06	6.00	Remove andtoss caps	Manual	5.76	11.52	0.00		2	0.00	7.92	3.60
7	4.07	7.00	Apply vaseline to connector fitting o-rings	Manual	3.60	3.60	0.00		1	0.00	0.00	3.60
8	4.08	8.00	Thread connector fittings into filter housings	Manual	5.04	10.08	0.00		2	10.08	0.00	0.00
9	4.06	9.00	Remove andtoss caps	Manual	5.76	11.52	0.00		2	0.00	7.92	3.60
10	4.07	10.00	Apply vaseline to connector fitting o-rings	Manual	3.60	3.60	0.00		1	0.00	0.00	3.60
11	4.08	11.00	Thread connector fittings into filter housings	Manual	5.04	10.08	0.00		2	10.08	0.00	0.00
12	4.09	12.00	Remove caps from connector fittings	Miscellaneous	5.76	5.76	0.00		1	0.00	0.00	5.76
13	4.10	13.00	Obtain air torque gun	Manual	1.80	1.80	0.00		1	0.00	0.00	1.80
14	4.11	14.00	Tighten connector fittings to filter housings using air torque gun	Manual	8.64	18.14	5.00		2	18.14	0.00	0.00
15	4.12	15.00	Replace caps on connector fittings	Manual	2.16	4.32	0.00		2	0.00	0.00	0.00
16	4.11	16.00	Tighten connector fittings to filter housings using air torque gun	Manual	8.64	18.14	5.00		2	18.14	0.00	0.00
17	4.12	17.00	Replace caps on connector fittings	Manual	2.16	0.00	0.00	16	2	0.00	0.00	0.00

Figure 10.5 Time study report generated by ProTime estimation module of the Proplanner.

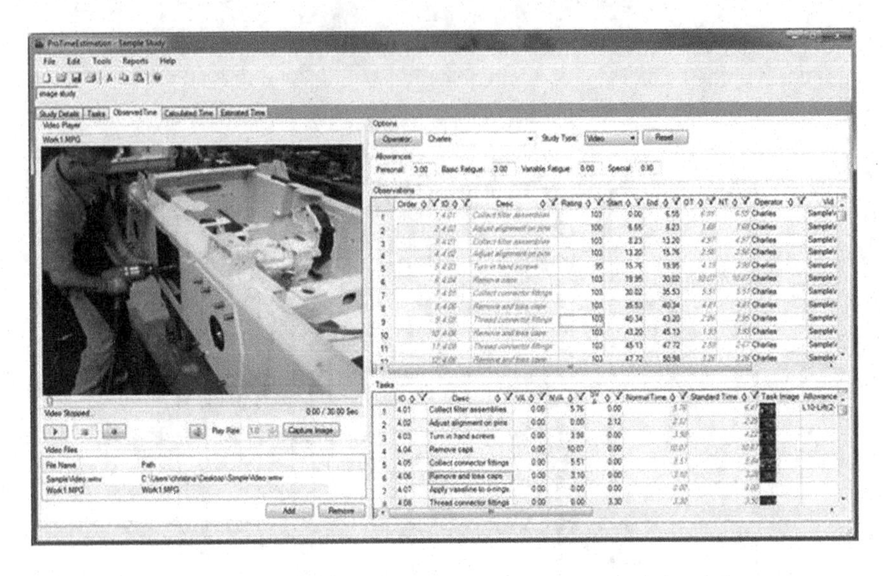

Figure 10.6 Video time study in ProTime estimation module of the Proplanner.

Figure 10.7 Method study: comparison of two methods of doing the same job by ProTime estimation module.

- Ergonomic analysis
- Lean analysis
- Man–machine charts
- Standard data libraries

The website can be accessed at www.acsco.com.

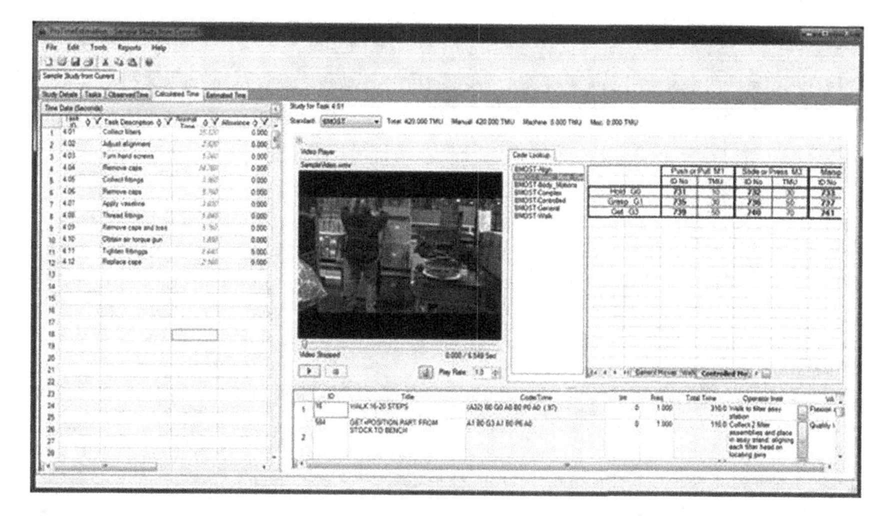

Figure 10.8 Basic MOST application in ProTime estimator module.

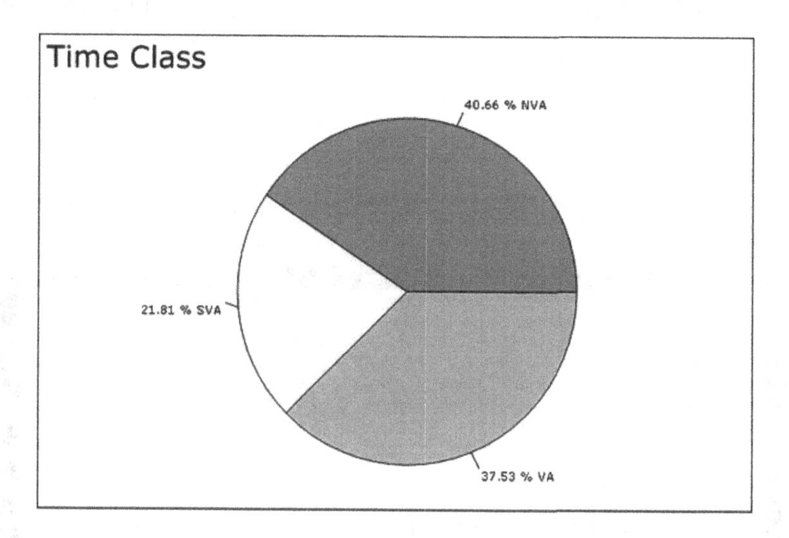

Figure 10.9 Breakdown of NVS/VA/SVA category times by ProTime estimation.

Figures 10.13 through 10.16 depict some products offered by Timer Pro Professional that can serve to facilitate work measurement.

10.3.3 Design Systems Inc.

Design systems Inc. is a consulting company that offers software solutions to perform utilization analysis, simulation services, manpower and dock optimization, and time and motion studies. It is a large engineering company employing a staff of more than 250 people. They provide program management, and

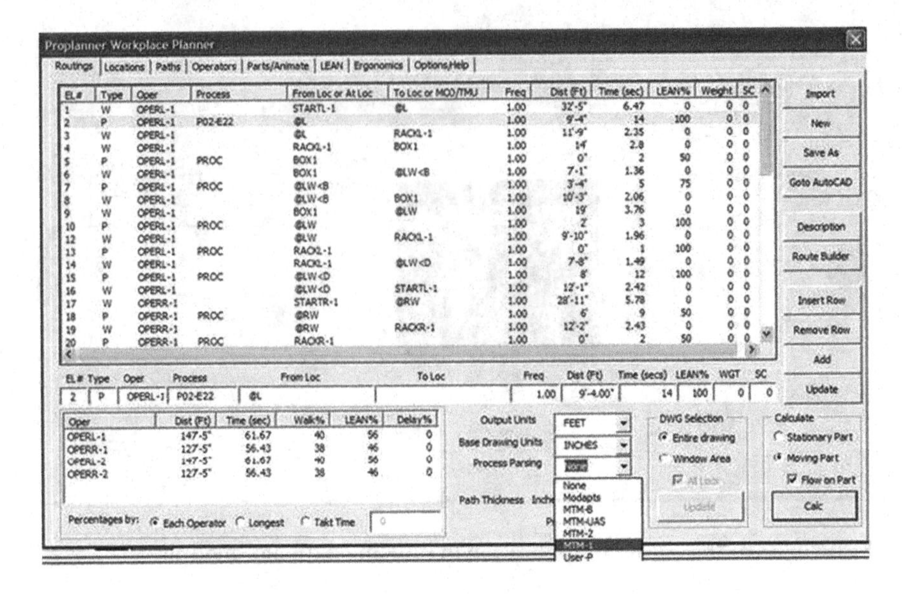

Figure 10.10 Utilization of PMTS data by workplace planner to enable facilities design.

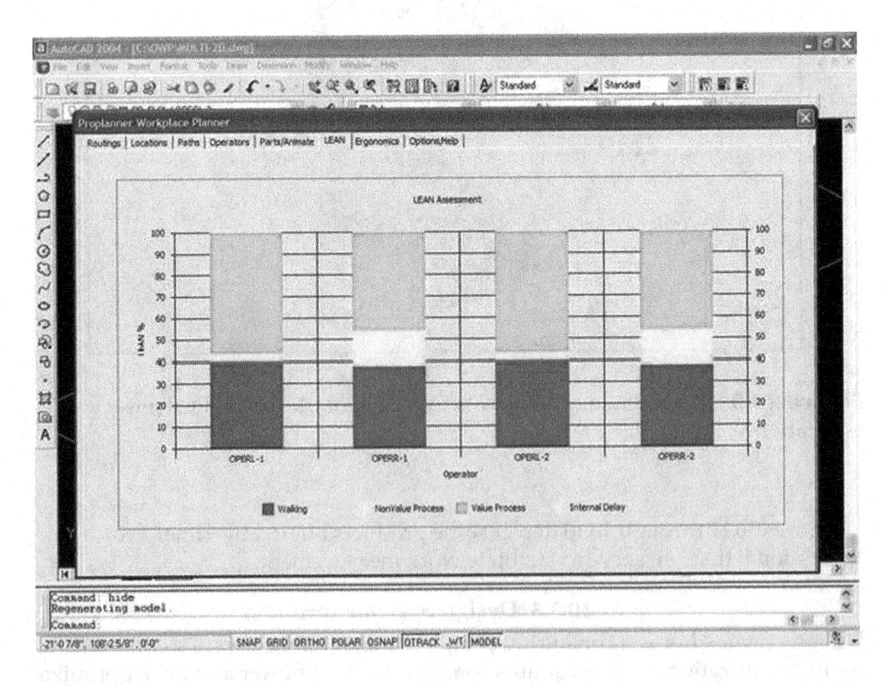

Figure 10.11 Calculation of value-added work content for workers computed by workplace planner.

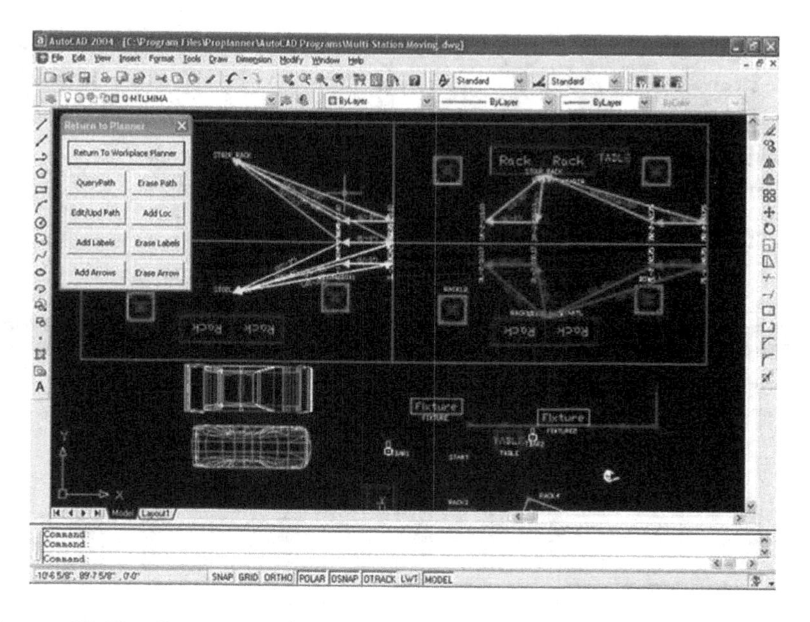

Figure 10.12 Generation of walk paths and evaluation of savings and overhead reductions by workplace planning module.

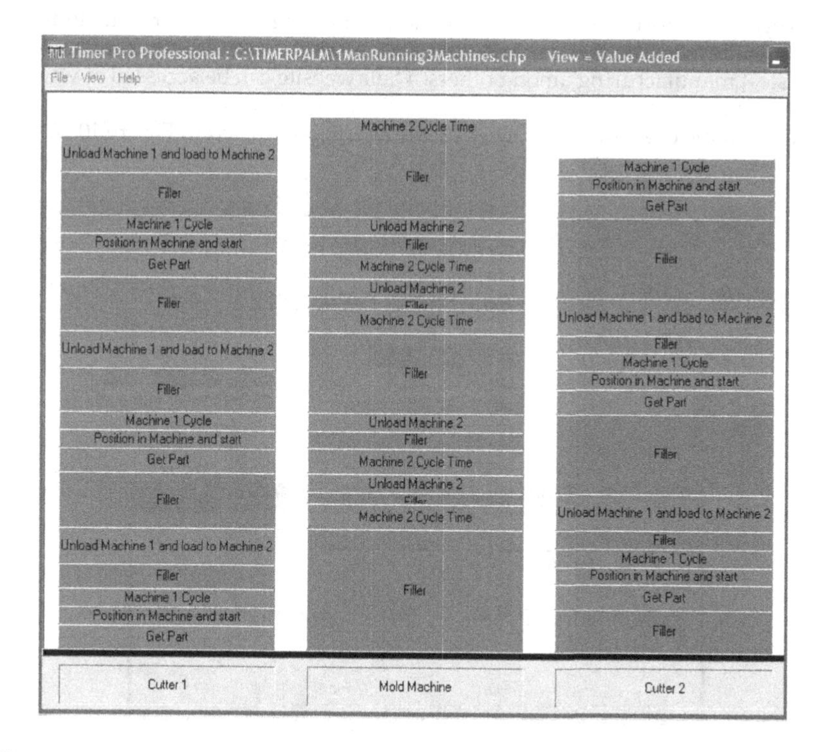

Figure 10.13 Man–machine chart for an operator operating three machines (Timer Pro).

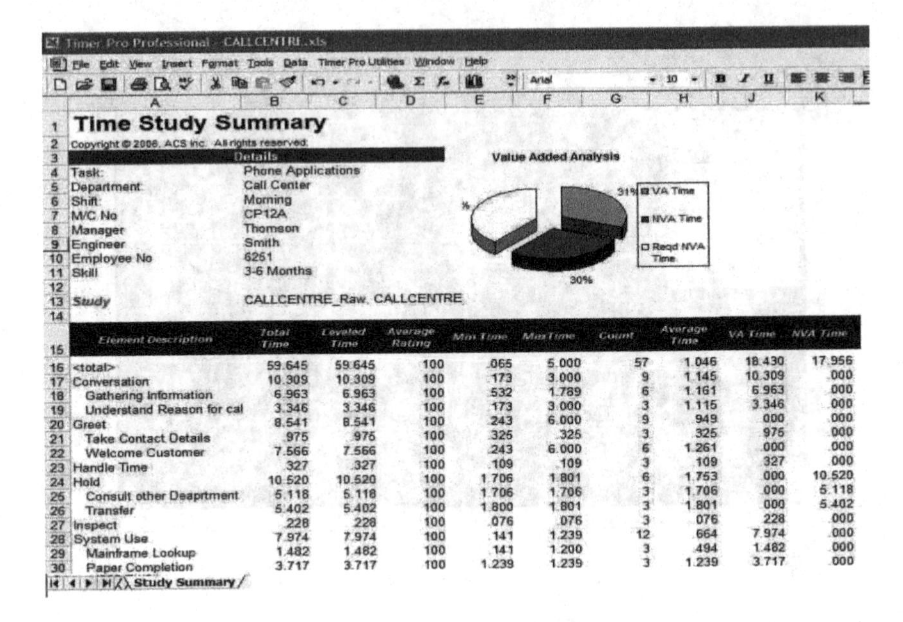

Element Description	Total Time	Leveled Time	Average Rating	Min Time	Max Time	Count	Average Time	VA Time	NVA Time
<total>	59.645	59.645	100	.065	5.000	57	1.046	18.430	17.956
Conversation	10.309	10.309	100	.173	3.000	9	1.145	10.309	.000
Gathering Information	6.963	6.963	100	.532	1.789	6	1.161	6.963	.000
Understand Reason for cal	3.346	3.346	100	.173	3.000	3	1.115	3.346	.000
Greet	8.541	8.541	100	.243	6.000	9	.949	.000	.000
Take Contact Details	.975	.975	100	.325	.325	3	.325	.975	.000
Welcome Customer	7.566	7.566	100	.243	6.000	6	1.261	.000	.000
Handle Time	.327	.327	100	.109	.109	3	.109	.327	.000
Hold	10.520	10.520	100	1.706	1.801	6	1.753	.000	10.520
Consult other Deaprtment	5.118	5.118	100	1.706	1.706	3	1.706	.000	5.118
Transfer	5.402	5.402	100	1.800	1.801	3	1.801	.000	5.402
Inspect	.228	.228	100	.076	.076	3	.076	.228	.000
System Use	7.974	7.974	100	.141	1.239	12	.664	7.974	.000
Mainframe Lookup	1.482	1.482	100	.141	1.200	3	.494	1.482	.000
Paper Completion	3.717	3.717	100	1.239	1.239	3	1.239	3.717	.000

Figure 10.14 Time study summary sheet (Timer Pro Professional).

conveyor and process engineering solutions for a range of industries including automobile, food and beverage pharmaceutical, package handling, health case, and steel manufacturing among others. Their website can be accessed at www.dsidc.com.

An example of a sample MODAPTS worksheet is depicted in Figure 10.17.

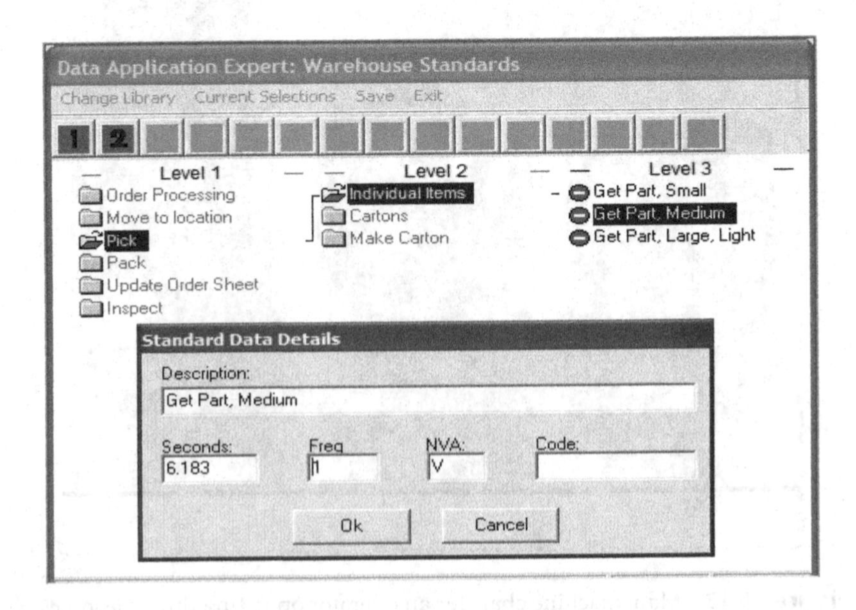

Figure 10.15 Standard data library (Timer Pro Professional).

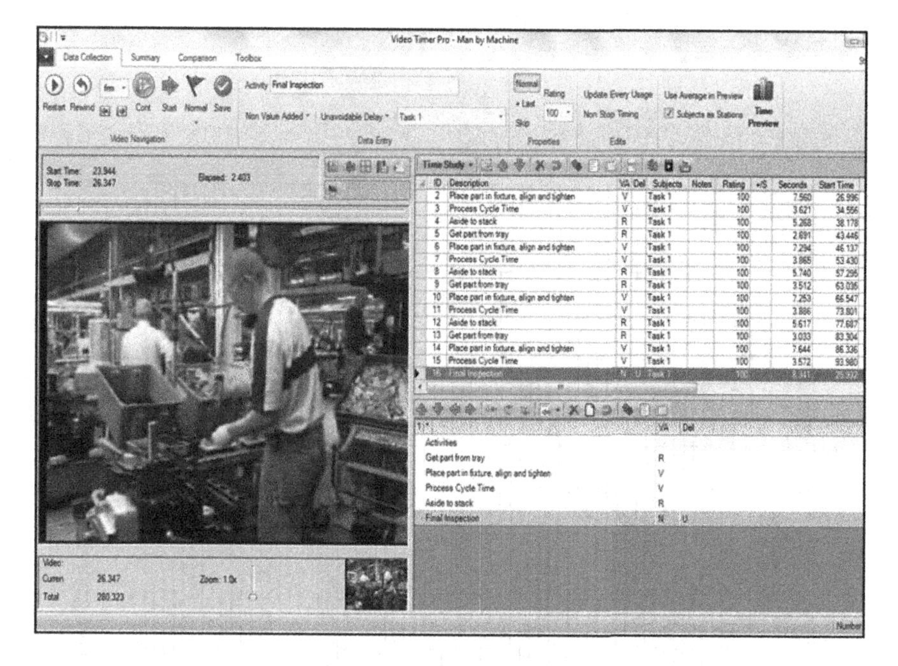

Figure 10.16 Video time and motion study by Video Timer Pro.

10.3.4 VRex Software Suite

The VRex software suite is offered by Office Line-Methods Line Industrial Engineering, a firm based in the Netherlands. VRex is a time study software that essentially teaches the user "how to" do a time study based on stopwatch time study methods and/or PMTS (we have discussed these in detail in Chapters 3 and 6, respectively). The software aims to set standard times faster, easier, and help the analyst make better-informed decisions. The company offers software solutions in the following formats:

- *Stopwatch time study:* This includes four modes: reference mode, film mode, tempo trainer mode, and measurement mode.
- *Time study:* This includes a video recording of the operation under consideration and analyzes it using the seven mudas of productivity.
- *Line balancing:* This includes the following features: real-time line balancing, workload constraints (line load and workstation load), variants and components, multiple balancing, benchmarking, and simulation.

Figures 10.18 through 10.23 depict some of the features of the stopwatch time study and PMTS-based time study solutions offered by this company. The website can be accessed at www.Officeline.be/Industrial_Engineering/methods_and_work_measurement.

10.4 NONTRADITIONAL METHODS OF MEASURING WORK

This section describes some of the instrumentation used for measuring physiological variables such as resting heart rate, oxygen consumption, aerobic capacity, and flicker fusion frequency as the nontraditional metrics of work.

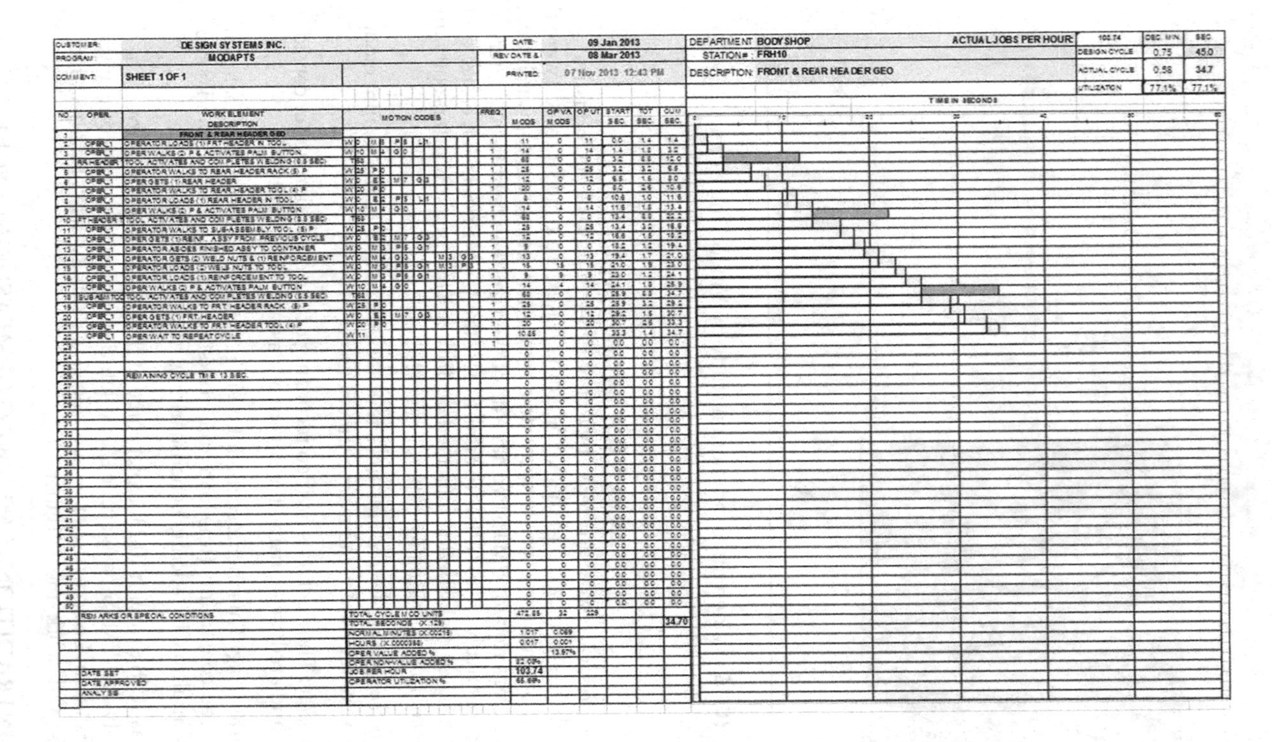

CUSTOMER:	DESIGN SYSTEMS INC.	DATE:	09 Jan 2013	DEPARTMENT: BODY SHOP	ACTUAL JOBS PER HOUR:	103.74	DEC. MIN	SEC
PROGRAM:	MODAPTS	REV DATE &	08 Mar 2013	STATION# : FRH10		DESIGN CYCLE	0.75	45.0
COMMENT:	SHEET 1 OF 1	PRINTED:	07 Nov 2013 12:43 PM	DESCRIPTION: FRONT & REAR HEADER GEO		ACTUAL CYCLE	0.58	34.7
						UTILIZATION	77.1%	77.1%

NO.	OPER.	WORK ELEMENT DESCRIPTION	MOTION CODES	FREQ.	MODS	OP VA MODS	OP UT	START SEC.	TOT SEC.	CUM SEC.
1		FRONT & REAR HEADER GEO								
2	OPER_1	OPERATOR LOADS (1) FRT HEADER IN TOOL	W0 M5 P5 L1	1	11	0	11	0.0	1.4	1.4
3	OPER_1	OPER WALKS (2) P & ACTIVATES PALM BUTTON	W10 M4 G0	1	14	0	14	1.4	1.8	3.2
4	RR HEADER	TOOL ACTIVATES AND COMPLETES WELDING (8.8 SEC)	T68	1	68	0	0	3.2	8.8	12.0
5	OPER_1	OPERATOR WALKS TO REAR HEADER RACK (5) P	W25 P0	1	25	0	25	3.2	3.2	6.5
6	OPER_1	OPER GETS (1) REAR HEADER	W0 E2 M7 G3	1	12	0	12	6.5	1.5	8.0
7	OPER_1	OPERATOR WALKS TO REAR HEADER TOOL (4) P	W20 P0	1	20	0	0	8.0	2.6	10.6
8	OPER_1	OPERATOR LOADS (1) REAR HEADER IN TOOL	W0 E2 P5 L1	1	8	0	8	10.6	1.0	11.6
9	OPER_1	OPER WALKS (2) P & ACTIVATES PALM BUTTON	W10 M4 G0	1	14	4	14	11.6	1.8	13.4
10	FT HEADER T	TOOL ACTIVATES AND COMPLETES WELDING (8.8 SEC)	T68	1	68	0	0	13.4	8.8	22.2
11	OPER_1	OPERATOR WALKS TO SUB-ASSEMBLY TOOL (5) P	W25 P0	1	25	0	25	13.4	3.2	16.6
12	OPER_1	OPER GETS (1) REINF. ASSY FROM PREVIOUS CYCLE	W0 E2 M7 G3	1	12	0	12	16.6	1.5	18.2
13	OPER_1	OPERATOR ASIDES FINISHED ASSY TO CONTAINER	W0 M3 P5 G1	1	9	0	0	18.2	1.2	19.4
14	OPER_1	OPERATOR GETS (2) WELD NUTS & (1) REINFORCEMENT	W0 M4 G3 M3 G3	1	13	0	13	19.4	1.7	21.0
15	OPER_1	OPERATOR LOADS (2) WELD NUTS TO TOOL	W0 M3 P5 G1 M3 P3	1	15	15	15	21.0	1.9	23.0
16	OPER_1	OPERATOR LOADS (1) REINFORCEMENT TO TOOL	W0 M3 P5 G1	1	9	9	9	23.0	1.2	24.1
17	OPER_1	OPER WALKS (2) P & ACTIVATES PALM BUTTON	W10 M4 G0	1	14	4	14	24.1	1.8	25.9
18	SUB ASM TOO	TOOL ACTIVATES AND COMPLETES WELDING (8.8 SEC)	T68	1	68	0	0	25.9	8.8	34.7
19	OPER_1	OPERATOR WALKS TO FRT HEADER RACK (5) P	W25 P0	1	25	0	25	25.9	3.2	29.2
20	OPER_1	OPER GETS (1) FRT HEADER	W0 E2 M7 G3	1	12	0	12	29.2	1.5	30.7
21	OPER_1	OPERATOR WALKS TO FRT HEADER TOOL (4) P	W20 P0	1	20	0	20	30.7	2.6	33.3
22	OPER_1	OPER WAIT TO REPEAT CYCLE	W11	1	10.65	0	0	33.3	1.4	34.7
23				1	0	0	0	0.0	0.0	0.0
24					0	0	0	0.0	0.0	0.0
25					0	0	0	0.0	0.0	0.0
26		REMAINING CYCLE TIME 13 SEC.			0	0	0	0.0	0.0	0.0
27					0	0	0	0.0	0.0	0.0
28					0	0	0	0.0	0.0	0.0
29					0	0	0	0.0	0.0	0.0
30					0	0	0	0.0	0.0	0.0
31					0	0	0	0.0	0.0	0.0
32					0	0	0	0.0	0.0	0.0
33					0	0	0	0.0	0.0	0.0
34					0	0	0	0.0	0.0	0.0
35					0	0	0	0.0	0.0	0.0
36					0	0	0	0.0	0.0	0.0
37					0	0	0	0.0	0.0	0.0
38					0	0	0	0.0	0.0	0.0
39					0	0	0	0.0	0.0	0.0
40					0	0	0	0.0	0.0	0.0
41					0	0	0	0.0	0.0	0.0
42					0	0	0	0.0	0.0	0.0
43					0	0	0	0.0	0.0	0.0
44					0	0	0	0.0	0.0	0.0
45					0	0	0	0.0	0.0	0.0
46					0	0	0	0.0	0.0	0.0
47					0	0	0	0.0	0.0	0.0
48					0	0	0	0.0	0.0	0.0
49					0	0	0	0.0	0.0	0.0
50					0	0	0	0.0	0.0	0.0

REMARKS OR SPECIAL CONDITIONS					
	TOTAL CYCLE MOD UNITS	472.65	32	229	
	TOTAL SECONDS (X .129)				34.70
	NORMAL MINUTES (X.00215)	1.017	0.069		
	HOURS (X.0000358)	0.017	0.001		
	OPER VALUE ADDED %		13.97%		
	OPER NON-VALUE ADDED %	82.03%			
DATE SET	JOB PER HOUR	103.74			
DATE APPROVED	OPERATOR UTILIZATION %	65.69%			
ANALYSIS					

Figure 10.17 Sample MODAPTS worksheet for a manufacturing operation (Design Systems Inc.).

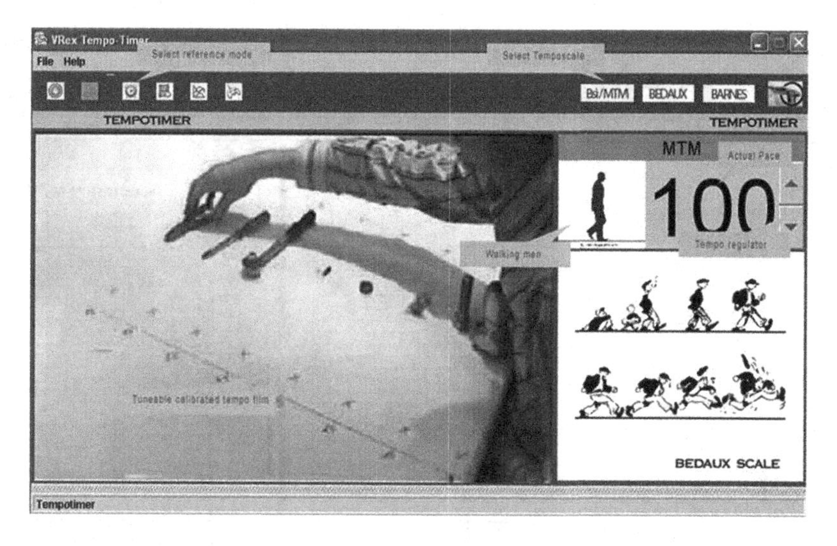

Figure 10.18 Reference mode of time study software offered by VRex.

Figure 10.19 Film mode of time study software offered by VRex.

As described in Chapter 9, specific instruments are used to measure each variable. For instance, aerobic capacity can be measured using a bicycle ergometer. Similarly other instrumentation such as the TEEM 100, COSMED K2, and Oxylog is used to measure oxygen consumption. All of these instruments are depicted in Figures 10.24 through 10.27. For specific details pertaining to pricing, refer to Chapter 9.

Bicycle ergometer: A bicycle ergometer is depicted in Figure 10.24. Several companies manufacture this product, but the basic design configuration is the same.

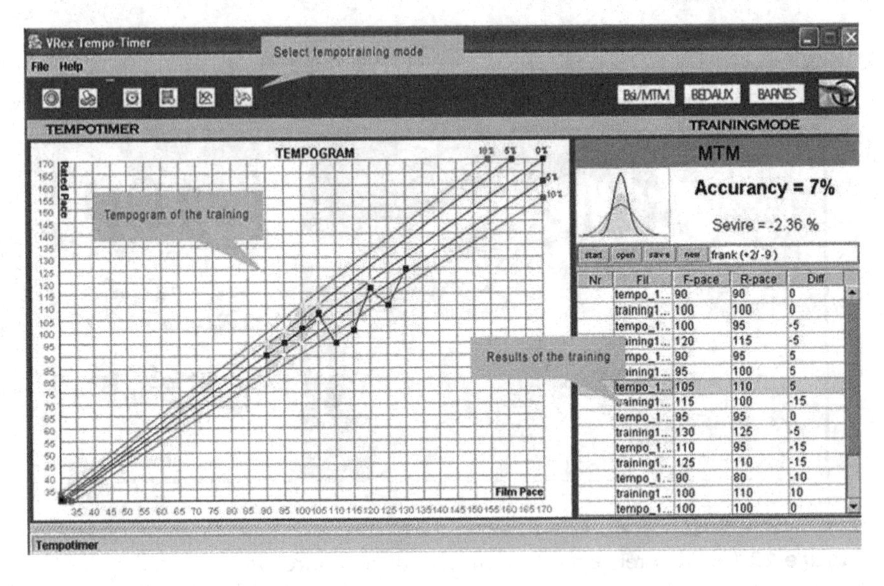

Figure 10.20 Tempotrainer mode of time study software offered by VRex.

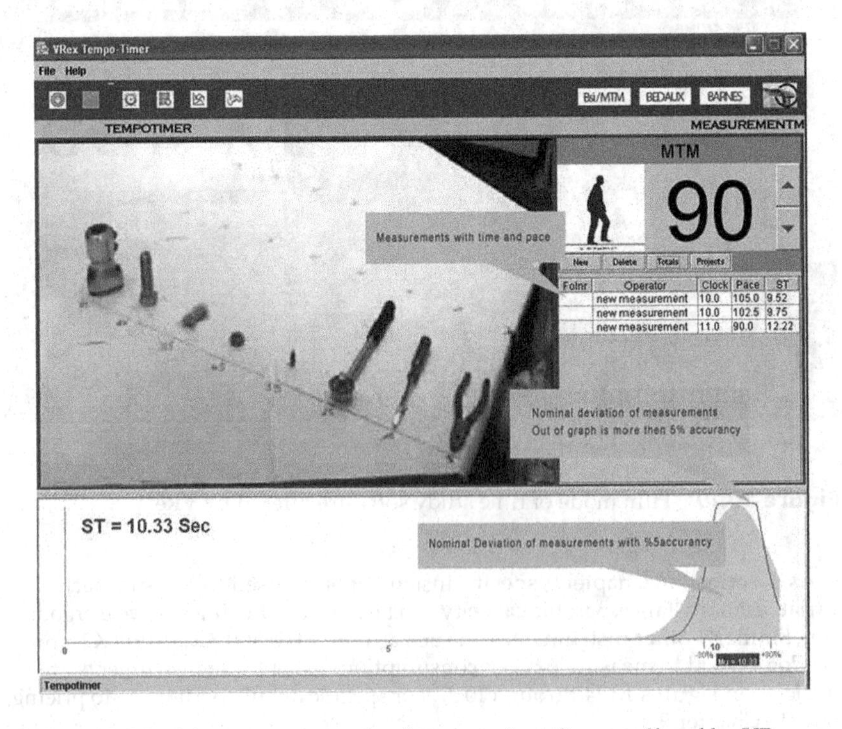

Figure 10.21 Measurement mode of time study software offered by VRex.

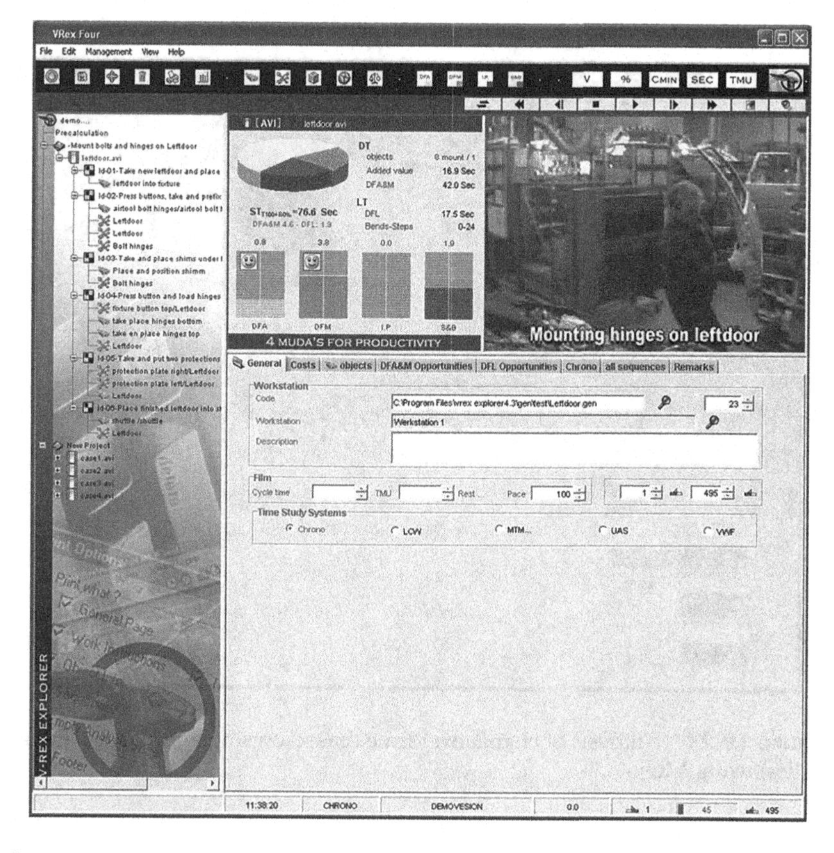

Figure 10.22 Time study analysis based on seven mudas of productivity (VRex Software).

It consists of a stationary bicycle. Different variables such as heart rate, oxygen consumption, aerobic capacity, etc. are recorded using relevant instrumentation.

TEEM 100: TEEM 100 is an electronic device used to perform metabolic analysis. TEEM is an acronym for total energy expenditure measurement and is manufactured by Aerosport Inc. It is depicted in Figure 10.25.

COSMED: The COSMED family of products is used for a variety of applications including pulmonary function testing, cardiopulmonary exercise testing, body composition, indirect calorimetry, etc. Their website can be accessed at www.cosmedusa.com. The COSMED K2 is an example of a device used for cardiopulmonary exercise testing. It is a wearable metabolic system that can be worn by athletes during actual events such as sprinting. The basic function is to measure metabolic activity as described in Chapter 9. Figure 10.26 depicts the COSMED K5.

Oxylog: The Oxylog is a device commonly used in mobile or transport ventilation. Thus, it is a machine that supports breathing. It is used to get oxygen into the lungs, remove carbon dioxide from the body, and in general to help the patient breathe easier. Drager manufactures it. The website can be accessed at www.draeger.com. An Oxylog is depicted in Figure 10.27.

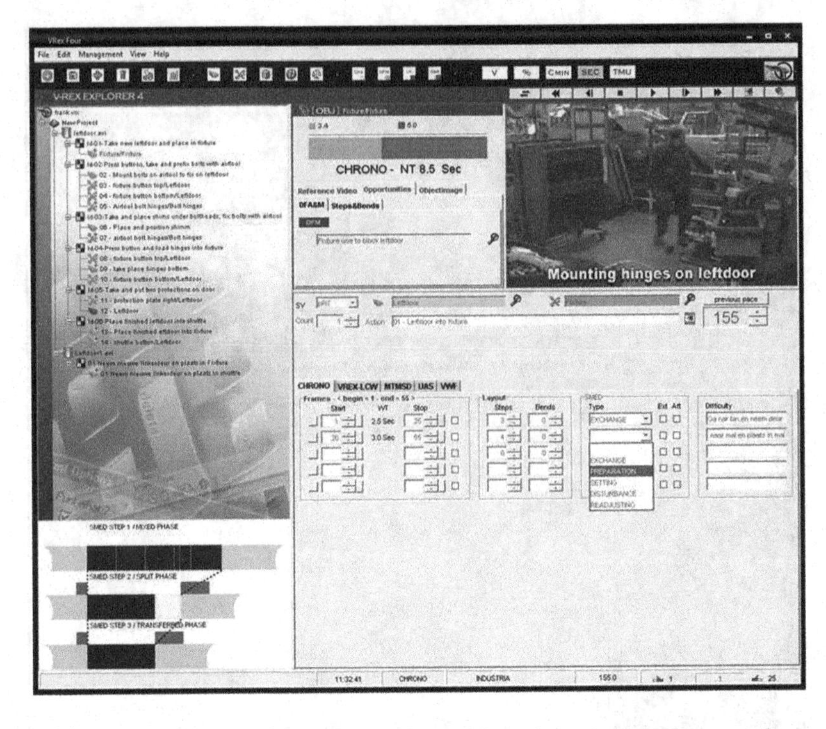

Figure 10.23 Analysis of changeover time (based on single-minute exchange of dies) using VRex.

Figure 10.24 Bicycle ergometer.

Figure 10.25 TEEM 100 metabolic analysis system.

Figure 10.26 The COSMED K5.

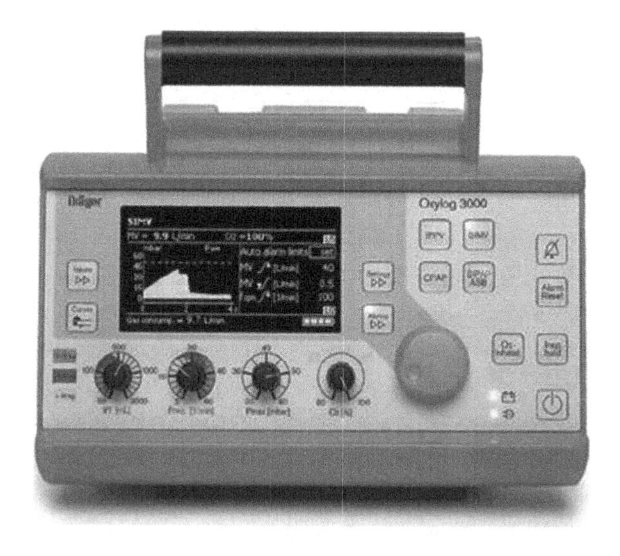

Figure 10.27 Oxylog.

10.5 SUMMARY

This chapter presented an overview of the "tools of the trade" used to conduct time and motion study within the larger context of work measurement. A variety of techniques to facilitate work measurement have been dealt with at length in this book. The tools that facilitate different elements of the technique, especially in the modern context were not presented until this chapter. Users and practitioners can gain additional information about the different software packages detailed in this chapter by accessing company websites. We hope that this chapter served as a valuable starting point to try and bring work measurement into the twenty-first century.

11 Using Work Measurement in Cost Estimation

11.1 INTRODUCTION

To be successful on the global market, product pricing must be competitive and must create revenue. Furthermore, products also must meet the expectations of consumers with regard to its functionality, quality, usability, reliability, esthetics, energy consumption, and environmental impact during its manufacture and operation and, finally, its disposal at the end of its useful life. These consumer expectations, along with legal and industry standards, have a crucial impact on the cost of a product. Before deciding on the selling price of a product, the manufacturer needs to estimate how much it will cost to produce and market the product.

Product pricing is heftily market dependent and is largely a function of supply and demand (e.g., the fluctuations in the current global price of oil). In a competitive market, company management may manipulate a product's price downwards in one of the two ways: (1) reduce the profit margin or (2) reduce its production cost. (A combination may be used, although it is expected that all efforts to minimize costs at the design stages have already been attempted. For instance, when due consideration has already been given to alternative materials, processes, quality requirements, and whatnot, at the design stages and have been thoroughly investigated.) Reducing the profit margin is not always a palatable option, unless the intent is to secure a greater market share or create a temporary increase in the market demand (assuming that the demand for the product is elastic), and is generally a short-term option. In the long term, a company must minimize its production costs without compromising on consumers' expectations. This means that the design of the product must be complete in terms of its ability to deliver the expected function without compromising its usability, quality, etc. Once the design of the product is complete in terms of its attributes, the manufacturer must forecast production quantity and determine the cost of manufacturing it accurately.

Cost estimating or *cost engineering* is the process of forecasting the costs of manufacturing a product (our discussion here pertains to the manufacturing industry, as opposed to the construction, mining, petrochemical, health, or some other industry). If the design of the product is complete and the production quantity is known, costs can be estimated accurately. Of course, the accuracy of the estimate is a function of time spent, and therefore the money spent, in generating the estimate.

In this chapter, our focus is on estimating costs associated with those aspects of manufacturing that rely on *job* or *time standards*. This limits our discussion to those activities, and associated costs, that are directly applicable to a product. Such activities typically include the following:

- Assembly
- Machining
- Inspection and testing
- Machine setup
- Tool adjustments (including tool changing)
- Rework

Before we discuss the methods to estimate costs that are dependent on time standards, we need to look at the various product cost components.

11.2 PRODUCT PRICING STRUCTURE

Figure 11.1 shows a typical product pricing structure. A brief description of each cost is given below:

Direct labor costs: These costs are associated with a specific product and include cost of all "hands-on" effort required to produce the product. As stated in the previous section, such costs include cost of effort associated with product assembly, machining, inspection, etc.

Direct material costs: This represents cost of all raw materials and all components procured from outside vendors that are included in the product. Cost of any material that does not form part of the product is not included.

Prime cost: It is the sum of direct labor cost and direct material cost.

Factory costs: These costs are also known as *indirect costs, overhead costs,* or *burden*. These are the total costs of items such as utilities (water, gas, and electricity), building rent, insurance, factory supplies, indirect factory labor (all labor costs that cannot be directly associated with producing the product), indirect factory materials (all material costs that cannot be associated with materials that form the product, for example, for testing and inspection).

Ex-factory cost: The total of prime cost and factory cost is called the ex-factory cost.

General costs: These costs include such costs as cost of engineering, purchasing, office supplies, depreciation, etc.

Manufacturing cost: The total of ex-factory cost and general cost is called the manufacturing cost.

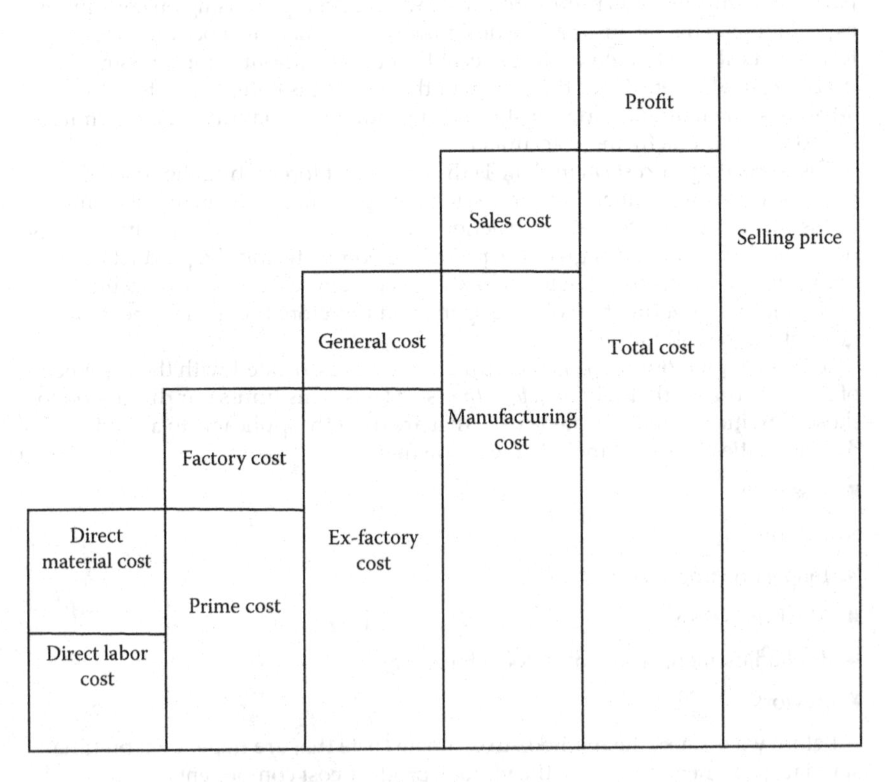

Figure 11.1 Components of product cost.

Sales costs: These costs include all costs associated with the "selling" function and include such costs as the cost of advertising, salary of sales personnel, storing the finished product, and shipping costs, etc.

Total cost: The sum of manufacturing cost and sales cost is the total cost.

Profit: It is the fraction of the total cost, over and above the total cost, that a company charges its customers. It is sometimes also called the profit margin.

Selling price: It is sum of total cost and profit.

11.3 ESTIMATING DIRECT LABOR COSTS

Direct labor costs are the only costs that require knowing time standards. Estimation of costs is generally required prior to the manufacture of a product. Sometimes cost estimates are also required for an existing product. Scenarios in which this typically happens include the following:

- Costs need to be updated to reflect current conditions (wage rate, inflation, etc.). Materials, process, equipment, etc. have undergone some changes.
- Product has undergone some design changes (e.g., new component design).
- There is a change in the geographical location of product manufacture.
- Accounting and costing procedures have changed.
- Product is marketed in distant markets.

Under such situations, time/unit needs to be reestablished. Procedures described in Chapter 3 can be used to estimate direct labor hours, and thereby direct labor costs, since the product is already being manufactured. In the majority of situations, however, cost estimates are required prior to manufacturing and direct labor hours cannot be estimated from direct time study. In such cases, PMTS described in Chapter 6 must be used to estimate direct labor hours.

Product manufacture requires procuring or manufacturing a product's components and assembling them. The costs of procured components are obtained from suppliers (vendors' selling price), but the costs of manufactured components and assembly must be estimated. The cost of estimating direct labor hours required to assemble a product requires experience and skill. The first step in the process is to generate a predecessor list showing the sequence of all assembly operations (A predecessor diagram shows the assembly starting stages on the extreme left and the final finished assembly stage on the extreme right). Table 11.1, for example, shows the predecessor list for a sample product assembly.

From Table 11.1, a predecessor diagram showing assembly stages can be prepared as shown in Figure 11.2. In Figure 11.2, operation 7 represents the finished assembly and operations 1, 3, 8, and 10 are the assembly starting operations.

For each assembly operation, a PMTS must be used to estimate completion time. The sum of all assembly operation times gives an estimate of direct labor hours required for assembly. For component manufacturing, machining times are estimated from certain formulas that are described later in this section. For each manufactured component, detailed process routing is required.

11.3.1 The Process Routing

A process routing primarily shows the sequence of manufacturing operations necessary to produce a component. It is typically in a columnar form and includes the following information:

- Sequence and description of operations
- Machine on which the operation is performed and the department name

Table 11.1: Predecessor List for a Sample Product Assembly

Assembly Task	Time/Unit (min)	Immediate Predecessor Task
1	2.0	–
2	1.0	1
3	0.7	–
4	1.0	3
5	1.5	2
6	1.0	4, 9
7	1.2	5, 6, 10
8	0.4	–
9	2.0	8
10	0.1	–
Total	10.9	

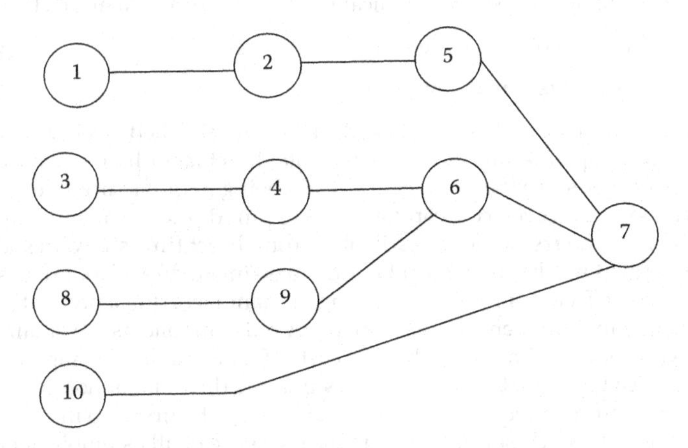

Figure 11.2 Predecessor diagram for the sample product assembly.

- Setup, handling, and run times for each operation (per piece)
- Shop order number corresponding to each operation
- Tools, jigs, and fixtures required for each operation (for nonroutine cases only)

A process routing sheet is required for each manufactured component. This is where the necessary experience and skills of the estimator come into play. Table 11.2 shows an example routing sheet of a sample component. The time it takes to machine each piece is multiplied by the production run (including scrap) to obtain the total machining time. The time for rework is added to this total.

11.3.2 Estimating Machining Times

A variety of equations is available in manufacturing engineering texts to estimate the machining time. Some of these equations are provided here (For detailed equations, the reader should consult a text on the manufacturing process or a manufacturing handbook that provides ways to determine machining times for a variety of machining operations).

Table 11.2: Process Routing Sheet for a Sample Component

Operation #	Description	Shop Order	Setup Time (h)	Handling Time (h)	Run Time (h)
1	Obtain material	A	0.25		
2	Setup machine	A	0.45		
3	Turn 1.5 in. OD	A		0.05	0.04
4	Turn 0.8 in. ID	A		0.05	0.03
5	Face two flanges	A		0.06	0.04
6	Cut off	A		0.02	0.02
	Total		0.70	0.18	0.13

Turning operations: For turning operations performed on a lathe, machining time is given as follows:

$$T = \frac{L}{f}$$

where T is the cutting time in minutes, L is the length of the cut in inches, and f is the feed rate in inches per revolution of the spindle.

Shaping operations: For shaping operations performed on a shaping machine, machining time is given as follows:

$$T = \frac{W}{N \times f}$$

where T is the cutting time in minutes, W the width of the piece to be cut in inches, N the number of machine strokes per minute required to generate a specified cutting speed (CS) in surface feet per minute, and f is the feed increment in inches per stroke. N is estimated from the following formula:

$$N = \frac{8CS}{L}$$

where L is the length of the cut + cutter over travel (clearance) at both ends.

Drilling and boring operations: Cutting time for drilling and boring operations is given by the following formula:

$$T = \frac{\text{Hole depth} + (1/2)\text{ hole diameter}}{f \times RPM}$$

where T is the cutting time in minutes; f is the feed rate in inches per revolution of the spindle; and RPM is the spindle speed in revolutions per minute.

Surface machining operations: For surface machining operations such as face milling or slab milling, the cutting time is given as follows:

$$T = \frac{L + D}{TT}$$

where T is the cutting time in minutes; L is the length of the cut in inches; and D is the cutter diameter in inches (for rough cuts, use D/2 in the above formula) and TT is the machine table travel in inches per minute. TT is given as follows:

$$TT = f_T \times N_T \times RPM$$

where f_T is the feed increment in inches per tooth; N_T is the number of teeth in the milling cutter; and RPM is the speed of the cutter in revolutions per minute.

If the cutting time is a function of the metal removal rate (MRR), that is the volume of material removed per minute, it can also be calculated.

MRR for turning operations is given by

$$\text{MRR} = \text{CS} \times d \times f$$

where CS is the cutting speed in surface feet per minute; d is the depth of cut in inches; and f is the feed rate in inches per spindle revolution.

MRR for shaping operations is given by

$$\text{MRR} = N \times f \times L \times d$$

where N is the number of machine strokes per minute; f is the feed increment in inches per stroke; L is the length of the stock removed per stroke in inches; and d is the depth of cut in inches.

MRR for drilling and boring operations is given by

$$\text{MRR} = \prod \times r^2 \times f \times \text{RPM}$$

where r is the hole radius in inches; f is the feed in inches per revolution of the spindle; and RPM is the spindle speed in revolutions per minute.

MRR for milling operations is given by

$$\text{MRR} = W \times d \times \text{TT}$$

where W is the width of cut in inches; d is the depth of cut in inches; and TT is the machine table travel time in inches per minute.

In machining operations, the cutting tool has to be replaced at the end of its life. Depending on the machining time, cutting tool may have to be replaced many times. How long will a cutting tool last and how long will it take to be replaced are significant time estimates in direct labor hour estimates. The life of a single-point cutting tool can be estimated by

$$V = C \times T^{-n}$$

where V is the CS in feet per minute and T is the tool life in minutes. The values of C, which is a constant, and n, which is the slope of the straight line describing the relationship between Log V and Log T, can be obtained from metal cutting handbook (one popular handbook has been developed by Metcut Research Inc. of Cincinnati, Ohio). Tool life for multiple point cutting tools can be estimated from metal cutting handbook as well.

The determination of tool replacement time and machine setup time for production run must be based on historical data and experience.

11.3.3 Direct Labor Costs

Once we have the time per piece (assembly or component machining), we can estimate the output per hour in terms of the number of pieces as follows:

$$\text{Pieces per hour} = \frac{60}{\text{Standard time in minutes}}$$

$$\text{Standard output per day} = [(\text{Pieces/hour}) \times 8 \text{ hours}] \text{ pieces}$$

The direct labor cost can be computed as follows:

Direct assembly labor cost/piece

= Wage rate ($/hour) × standard assembly time (hour/piece)

Direct assembly labor cost/day

= (#pieces/day)[standard time (hour/piece)] × wage rate ($/hour)

Direct machining labor cost/piece

= Sum of [standard time (hour/component)] for all components

It should be noted that the wage rate may or may not include the cost of fringe benefits. If such cost is not included, the wage rate must be modified to include it. The actual wage rate in that case is

$$\text{Actual wage rate} = \text{Wage rate} \times (1 + F + U + W + H)$$

where F is the FICA fraction; U is the unemployment compensation fraction; W is the workers' compensation fraction; and H is the health and other insurance compensation fraction.

11.4 SUMMARY

In this chapter, we have looked at that aspect of the product cost component that requires time standards for estimation, namely the direct labor cost. Indirect labor costs that permit measurement of time/unit of indirect labor exertions can be estimated similarly. For estimating costs associated with other indirect labor exertions, first the time of such exertions for each unit needs to be determined. Some of the techniques discussed in Chapters 2 and 10 can be used for this purpose. Estimation of product cost is a complex and tedious process and this chapter, by no means, addresses that complexity. Rather, it shows the critical role time standards play in the estimation of the overall product cost. As Figure 11.1 shows product cost has many components. To estimate all these components, one must rely on manufacturing cost estimation experts. We recommend that, at the very least, the reader should refer to the cost estimation reference provided in the reading list.

Suggested Reading

S.A. Konz and S. Johnson, *Work Design: Industrial Ergonomics*, 5th edition, Holcomb Hathaway, 2000.

A. Mital, A. Desai, A. Subramanian, and A. Mital, *Product Development: A Structured Approach to Consumer Product Development, Design, and Manufacture*, 2nd edition, Elsevier, 2014.

B. Niebel and A. Freivalds, *Methods, Standards & Work Design*, 10th edition, WCB/McGraw-Hill, 1999.

P.F. Ostwald, *Engineering Cost Estimating*, 3rd edition, Prentice-Hall, 1991.

D.R. Sule, *Manufacturing Facilities: Location, Planning, and Design*, 3rd edition, CRC Press, 2009.

Appendix: Source Listing of Rest Allowance Determination Program Based on Metabolic Energy Expenditure

```
10'   PROGRAM TO COMPUTE THE REST PERIOD AS A PERCENT OF TOTAL
      SHIFT PERIOD BY
20'   ANIL MITAL
30'   UNIVERSITY OF CINCINNATI
40'
50'
60'   THIS PROGRAM COMPUTES THE REST PERIOD AS A PERCENT OF TOTAL
      SHIFT DURATION
70'   TWO LOOKUP TABLES ARE IMPLEMENTED IN THE PROGRAM GIVEN THE
      AGE AND
80'   AEROBIC CAPACITY OF THE WORKER, IT COMPUTES THE PHYSIOLOGICAL
      CONDITION
90'   WHICH COULD FALL IN FIVE CATEGORIES
100'
110'  SECOND TABLE COMPUTES TOTAL AEROBIC CAPACITY FOR 18 YEAR OLD
      MALE OR FEMALE
120'  FOR 24 HOURS WHEN THE PHYSIOLOGICAL CONDITION IS KNOWN
130'
140'
150   DIM T(20), E(20), MX(5), FX(5)
160   DIM WCAP1(5), WCAP2(5), WCAP(3), WCAP4(5), WCAP(5)
170   DIM MCAP1(5), MCAP2(5), MCAP3(5), MCAP4(5), MCAP5(5)
180'
190'
200'  THE FOLLOWING IS THE TABLE FOR 18 YEAR OLD MALE (MX) AND
      FEMALE(FX)
210   MX(1)=2100: MX(2)=2575: MX(3)=3050: MX(4)=3525: MX(5)=4000
220   FX(1)=2100: FX(2)=2325: FX(3)=2550: FX(4)=2775: FX(5)=3000
230'
240'
250'  TABULAR VALUES FOR WOMEN'S PHYSIOLOGICAL CONDITION DEPENDING
      UPON AGE
260'  AND AEROBIC CAPACITY
270'
280   WCAP1(1)=24: WCAP1(2)=30: WCAP1(3)=37: WCAP1(4)=48"
      WCAP1(5)=100
290   WCAP2(1)=20: WCAP2(2)=27: WCAP2(3)=33: WCAP2(4)=44:
      WCAP2(5)=100
300   WCAP3(1)=17: WCAP3(2)=23: WCAP3(3)=30: WCAP3(4):=41:
      WCAP3(5)=100
310   WCAP4(1)=15: WCAP4(2)=20: WCAP4(3)=27: WCAP4(4)=37:
      WCAP4(5)=100
320   WCAP5(1)=13: WCAP5(2)=17: WCAP5(3)=23: WCAP5(4)=34:
      WCAP5(5)=100
330'
340'
350'  TABULAR VALUES FOR MEN'S PHYSIOLOGICAL CONDITION DEPENDING
      UPON AGE
360'  AND AEROBIC CAPACITY
370'
380   MCAP1(1)=25: MCAP1(2)=33: MCAP1(3)=42: MCAP1(4)=52:
      MCAP1(5)=100
390   MCAP2(1)=23: MCAP2(2)=30: MCAP2(3)=33: MCAP2(4)=49:
      MCAP2(5)=100
```

```
400   MCAP3(1)=20: MCAP3(2)=26: MCAP3(3)=35: MCAP3(4)=44:
      MCAP3(5)=100
410   MCAP4(1)=18: MCAP4(2)=24: MCAP4(3)=33: MCAP4(4)=42:
      MCAP4(5)=100
420   MCAP5(1)=16: MCAP5(2)=22: MCAP5(3)=30: MCAP5(4)=40:
      MCAP5(5)=100
430'
440'
450   AGE1=29: AGE2=39: AGE3=49: AGE4=59: AGE5=69
460'
470'
480   INPUT "DATE: ", DATE$
490   INPUT "JOB NUMBER: ", JOB$
500'
510   INPUT "JOB CODE: ", JOB CODE$
520   INPUT "WORKERNAME: ", W. NAME$
530   INPUT "SEX (M-MALE", F-FEMALE): ", SEX$
540   IF SEX$="M" OR SEX$="F" THEN GOTO 570
550   GOTO 530
560'
570   INPUT "AGE (BETWEEN 20 AND 69): ", AGE
580   IF AGE<19 OR AGE>69 THEN GOTO 570
590'
600   INPUT "WEIGHT IN POUNDS: ", WEIGHT
610   IF WEIGHT<1 THEN GOTO 600
620'
630'
640   INPUT "IS AEROBIC CAPACITY KNOWN? (Y/N): ", AC
650   IF AC YESNO$="N" THEN GOTO 1260
660   INPUT "ENTER AEROBIC CAPACITY IN ML/KG/MIN (1-100): ",
      ACAPACITY
670'
680'
690'  SINCE AEROBIC CAPACITY IS KNOWN, PHYSIOLOGICAL CONDITION IS
      COMPUTED
700'  FROM THE TABLE
710   IF SEX$="M" GOTO 1000
720'
730'
740'  THIS SECTION COMPUTES PHYSIOLOGICAL CONDITION OF WOMEN
750'
760   IF AGE>AGE1 THEN GOTO 800
770   FOR I=1 TO 5
780   IF ACAPACITY<WCAP1(I) THEN GOTO 1210
790   NEXT I
800   IF AGE>AGE2 THEN GOTO 840
810   FOR I=1 TO 5
820   IF ACAPACITY<WCAP2(I) THEN GOTO 1210
830   NEXT I
840   IF AGE>AGE3 THEN GOTO 880
850   FOR I=1 TO 5
860   IF ACAPACITY<WCAP3(I) THEN GOTO 1210
870   NEXT I
880   IF AGE>AGE4 THEN GOTO 920
890   FOR I=1 TO 5
900   IF ACAPACITY<WCAP4(I) THEN GOTO 1210
910   NEXT I
920   FOR I=1 TO 5
930   IF ACAPACITY<WCAP5(I) THEN GOTO 1210
```

```
940   NEXT I
950'
960'
970'
980'  THIS SECTION COMPUTES PHYSIOLOGICAL CONDITION OF MEN
990'
1000  IF AGE>AGE1 THEN GOTO 1040
1010  FOR I=1 TO 5
1020  IF ACAPACITY<MCAP1(I) THEN GOTO 1210
1030  NEXT I
1040  IF AGE>AGE2 THEN GOTO 1080
1050  FOR I=1 TO 5
1060  IF ACAPACITY<MCAP2(I) THEN GOTO 1210
1070  NEXT I
1080  IF AGE>AGE3 THEN GOTO 1120
1090  FOR I=1 TO 5
1100  IF ACAPACITY<MCAP3(I) THEN GOTO 1210
1110  NEXT I
1120  IF AGE>AGE4 THEN GOTO 1160
1130  FOR I=1 TO 5
1140  IF ACAPACITY<MCAP4(I) THEN GOTO 1210
1150  NEXT I
1160  FOR I=1 TO 5
1170  IF ACAPACITY<MCAP5(I) THEN GOTO 1210
1180  NEXT I
1190'
1200'
1210  PHYS. COND=I
1220  PRINT "COMPUTED PHYSIOLOGICAL CONDITION: ", PHYS. COND.
1230  GOTO 1310
1240'
1250'
1260  PRINT "SINCE AEROBIC CAPACITY IS NOT KNOWN, ENTER ESTIMATED"
1270  INPUT "PHYSIOLOGICAL CONDITION (1=LOW, 2=AVERAGE, 3=HIGH)",
      PHYS. COND.
1280  IF PHYS. COND<1 OR PHYS. COND>3 THEN GOTO 1270
1290'
1300'
1310  INPUT "HOURS OF SLEEP PER DAY (Y1): ", Y1
1320  INPUT "SHIFT DURATION IN HOURS (Y2): ", Y2
1340'
1350'
1360' ENERGY AVAILABLE FOR 18 YEAR OLD IS COMPUTED FROM THE TABLE
1370  IF SEX$="F" THEN GOTO 1400
1380  X=MX (PHYS. COND.)
1390  GOTO 1410
1400  X=FX (PHYS. COND.)
1410  PRINT ËNERGY AVAILABLE TO 18 YEAR OLD: ", X
1420'
1430'
1440' ADJUSTING FOR AGE - 1% DECLINE
1450  LET X2=X-(AGE-18)xXx.01
1460'
1470'
1480' ADJUST X1 FOR THE INFLUENCE OF FOOD
1490  LET X2=.9Xx1
1500'
1510'
1520' CALCULATE BASAL METABOLISM DURING SLEEP
```

```
1530  LET X3=(.9xWEIGHTxY1)/2.2
1540'
1550'
1560' CALCULATE X4, THE ENERGY REQUIREMENT WHEN NOT SLEEPING OR
      WHEN NOT
1570' AT WORK
1580  INPUT "IS ENERGY (KCAL/MIN) WHEN NOT SLEEPING/WORKING KNOWN?:",
      ER.YESNO$
1590  IF ER. YESNO$="Y"THEN INPUT "ENERGY REQUIRED": ", ER
1600  IF ER. YESNO$ < > "Y"THEN GOTO 1630
1610  LET X4=Y3x60xER
1620  GOTO 1670
1630  LET X4=Y3x60x1.5
1640'
1650'
1660' DETERMINE TOTAL ENERGY AVAILABLE FOR WORK (X5)
1670  LET X5=X2-X3-X4
1680  PRINT "X5=", X5
1690'
1700'
1710  INPUT "NUMBER OF JOBS N: ", N
1720  LET X6=0
1730  TT=0
1740  FOR I=1 TO N
1750  INPUT "ENTER TIME DURATION IN HOURS FOR THE ITH TASK: ", T(I)
1760  INPUT "ENTER TOTAL ENERGY REQUIRED/HOUR FOR THE ITH TASK: ",
      E(I)
1770  LET X6=X6+T(I)xE(I)
1780  LET TT=TT+T(I)
1790  NEXT I
1800'
1810'
1820  IF TT=Y2 THEN GOTO 1860
1830  PRINT "TOTAL TIME DURATION (HOURS) IS NOT EQUAL TO Y2 (SHIFT
      DURATION) CHECK INPUT"
1840  GOTO 640
1850'
1860  IF X5>X6 THEN R=0 ELSE R=(X6/X5-1)x100
1870  PRINT "REST PERIOD AS A PERCENTAGE OF SHIFT DURATION: ", R
1880'
1890  INPUT "DO YOU WANT A HARD COPY? (Y/N): ", HARD. YESNO$
1900  IF HARD YESNO$ < > "Y"GO TO 2280
1910'
1920  LPRINT
1930  LPRINT "PROGRAM OUTPUT"
1940  LPRINT
1941  LPRINT
1950  LPRINT "DATE: ", DATE$
1951  LPRINT
1960  LPRINT "JOB NUMBER: ", JOB$
1961  LPRINT
1970  LPRINT "JOB CODE: ", JOB.CODE$
1971  LPRINT
1980  LPRINT "WORKER NAME: ", W.NAME$
1981  LPRINT
1990  LPRINT "SEX: ", SEX$
1991  LPRINT
2000  LPRINT "AGE(YEARS): ", AGE
2001  LPRINT
```

```
2010  LPRINT "WEIGHT IN POUNDS: ", WEIGHT
2011  LPRINT
2020'
2030  IF AC. YESNO$="N" GOTO 2070
2040  LPRINT "AEROBIC CAPACITY IN ML/KG/MIN: ", ACAPACITY
2050  LPRINT "PHYSIOLOGICAL CONDITION COMPUTED FROM THE TABLE: ",
      PHYS COND
2060  GOTO 2080
2070  LPRINT "ESTIMATED PHYSIOLOGICAL CONDITION: ", PHYS. COND
2080'
2081  LPRINT
2090  LPRINT "HOURS OF SLEEP PER DAY: ", Y1
2100  LPRINT "SHIFT DURATION: ", Y2
2110'
2111  LPRINT
2120  PRINT "ENERGY AVAILABLE FOR 18 YEAR OLD FROM THE TABLE (X),
      IN KCAL: ", X
2130  LPRINT "ENERGY REQUIRED WHEN NOT SLEEPING AND NOT WORKING,
      IN KCAL: ", X4
2140  LPRINT "TOTAL ENERGY (KCAL) AVAILABLE FOR WORK: ", X5
2150  LPRINT
2151  LPRINT
2160  LPRINT "TOTAL NUMBER OF JOBS: ", N
2170  LPRINT
2180  LPRINT "TIME DURATION (IN HOURS) OF JOB      ENERGY REQUIRES
      (KCAL/HOUR) FOR JOB"
2190  FOR I=1 TO N
2200  LPRINT T(I)
2210  NEXT I
2220'
2221  LPRINT
2230  LPRINT
2240  LPRINT "TOTAL ENERGY REQUIRED FOR JOB (KCAL): ", X6
2250  LPRINT "REST PERIOD AS A PERCENT OF SHIFT DURATION: ", R
2260'
2270'
2280  INPUT "DO YOU WANT TO RUN THE NEXT JOB? (Y/N), JOB.YESNO$
2290  IF JOB.YESNO$="Y" GOTO 480
2300  END
```

Index

F

G

H

I

J

M

N

O

P

Q

V

W

Printed in the United States
by Baker & Taylor Publisher Services